# ADVANCED COMMON CORE MATH EXPLORATIONS

**GRADES 5–8**

## MATH EXPLORATIONS

## *Ratios, Proportions,& Similarity*

JERRY BURKHART

**Routledge**
Taylor & Francis Group

NEW YORK AND LONDON

First published in 2016 by Prufrock Press Inc.

Published 2021 by Routledge
605 Third Avenue, New York, NY 10017
2 Park Square, Milton Park, Abingdon, Oxon OX14 4RN

*Routledge is an imprint of the Taylor & Francis Group, an informa business*

Copyright © 2016 by Taylor & Francis Group

Cover design by Raquel Trevino and layout design by Allegra Denbo

ISBN 13: 978-1-0321-4435-1 (hbk)
ISBN 13: 978-1-6182-1529-1 (pbk)

DOI: 10.4324/9781003232797

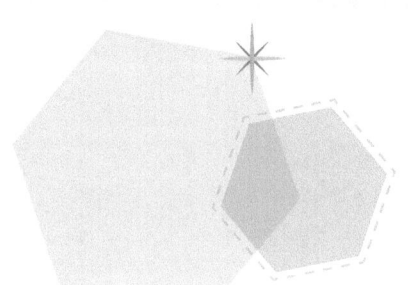

# Table of Contents

Preface................................................................ vii
A Note to Students.............................................. ix
A Note to Teachers............................................. xi
Introduction ......................................................1
Connections to the Common Core State Standards.....................3
Teacher's Guide...................................................7

**Exploration 1:** The Incredible Shrinking Universe............... 19

**Exploration 2:** Ramps, Paints, and Hot Air Balloons........... 41

**Exploration 3:** Gear Up!...................................... 71

**Exploration 4:** Perplexing Percentages....................... 89

**Exploration 5:** Scaling a Tower ............................. 125

**Exploration 6:** Keep It in Proportion........................ 145

**Exploration 7:** Grab Bag..................................... 167

**Exploration 8:** Expanding and Contracting.................... 191

**Exploration 9:** Pythagorean Connections..................... 223

**Exploration 10:** Twist and Shrink........................... 237

References ...................................................... 253
About the Author ............................................... 255
Common Core State Standards Alignment .......................... 257

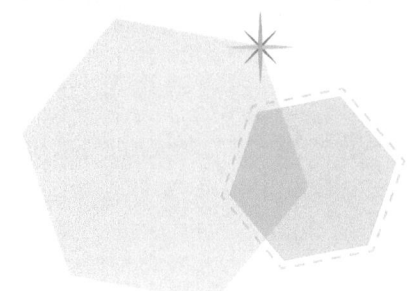

# Preface

Many people have played key roles in the books in this series. I would first like to thank Sarah Scott, without whose passion, support, and imagination the books would not exist. My daughter, Annie, has greatly improved my writing through many hours of careful reading, insightful suggestions, and lively discussion. Sue Wygant has offered numerous ideas to improve the usability of the activities and to help them better support the inquiry-based approach that they represent.

If you have purchased other books in this series, you will notice some changes in this book! Ian Byrd of byrdseed.tv has provided wonderful feedback on the format, resulting in a more appealing, open appearance and a more informal, streamlined writing style. I am grateful for his willingness to share his expertise, and I am excited for the potential to make challenging mathematical content more broadly accessible.

I would also like to acknowledge the staff at Prufrock Press. I appreciate their investment in and support of such a large undertaking. Lacy Compton has been wonderfully patient and skilled in balancing the requirements of her job as editor with respecting my vision and purpose as an author.

Finally, I would like to thank my students over the years. Their love of learning has been an inspiration to me, and these books are full to the brim with their ideas!

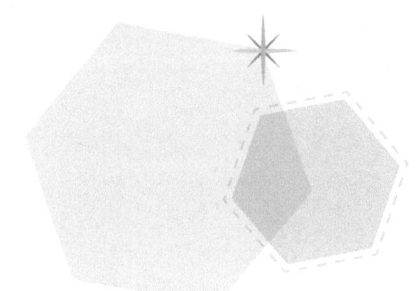

# A Note to Students

Welcome, math explorers! You are about to embark on an adventure in learning. As you navigate the mathematical terrain in these activities, you will discover that "doing the math" means much more than calculating quickly and accurately. It means using your creativity and insight to question, investigate, describe, analyze, predict, and prove. It means venturing into unfamiliar territory, taking risks, and finding a way forward when you are not sure which direction to go. And it means making discoveries that will expand your mathematical imagination.

Of course, the job of an explorer is hard work. At times, it will take a real effort on your part to keep going. You may spend days or more pondering a single question. Sometimes, you might even get completely lost. The process can be demanding—but it is also rewarding. There is nothing quite like the feeling of making a breakthrough after a long stretch of hard work, and seeing a whole new world of ideas open up before your eyes!

These explorations are challenging, so you might want to team up with a partner or two on your travels—to discuss plans and strategies and to share the rewards of your hard work. Even if you don't always reach your final destination, you will find that the journey was worth taking. So, gear up for some hard work and adventure . . . and start exploring!

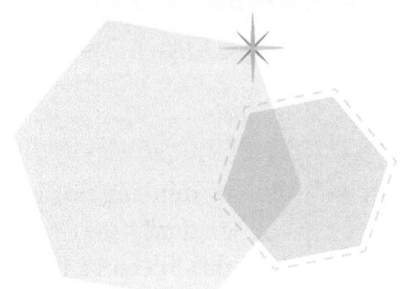

# A Note to Teachers

When teachers see these explorations for the first time, some believe them to be too challenging. My experience has been different. I believe that we tend to underestimate our students. It is true that they will not typically be able to complete every problem in an activity. They are unlikely to answer most questions perfectly. Certainly, they will not finish quickly. They will sometimes struggle, and they may need your support during that process. But your students *can* be successful with the explorations—and so can you!

Often, when our students struggle, especially in math, we tend to believe that our job is to fix the "problem" by explaining more clearly or in a different way. But, in fact, it is when our students are struggling with mathematical ideas that there is the greatest potential for deep learning. Our job is not so much to explain as to ensure that their struggle is productive.

Your keys to success in this endeavor are patience, curiosity, an open mind, and a certain level of trust (perhaps even a leap of faith!) in yourself, your students, and the activities. Building this type of deep learning into your teaching practice takes time. Think of it as a *process* of learning and growing along with your students. Do not take on too much at the beginning. Assign fewer problems, and listen closely as your students talk and write about their ideas. As you reflect and come to understand how they are thinking about the problems, you will learn more about the math and about how to facilitate conversations that help students become authors of their own learning.

In the end, it is the *thinking* that creates the real learning. Talented students (really, *all* students) who develop a habit of thinking hard about challenging problems and ideas grow their mathematical capacity in deep and powerful ways. The times when they hit a wall and feel that they are making no progress are probably the times they are learning the most. Time and time again, my students come to class bursting with excitement, telling me, for example, that they were riding the bus home when the answer to a question they had been thinking about for days suddenly came to them. I love it when this happens, because they are able to see that their brains were engaged with the problem even when they were not aware of it! But the breakthrough occurs only because of the work they did while they felt stuck.

As powerful as it is, this way of learning can be "messy." Ideas do not come tied up in neat packages. You and your students gradually learn to live with ambiguity as you consider multiple strategies and make connections between different ways of thinking about problems. You begin to see that deep understanding develops gradually over time as your mind knits together various strands of ideas. In fact, this tolerance for ambiguity is a chief characteristic of successful mathematicians! The presence of uncertainty and confusion as you work toward understanding is an inescapable and wonderful part of the process of doing mathematics.

That is not to say that you normally leave things in this state of uncertainty. When students are working toward specific learning goals, it is important in the end to organize, clarify, and summarize what was learned. But this happens *after* students have wrestled with the ideas and *at the level* that they are prepared to make sense of them. The "debriefing" strategy described in the Eight Motivation Strategies section of the Teacher's Guide will support your work in this process.

Teachers as well as students begin these activities with different levels of comfort and confidence. Some teachers do not think of themselves as "math people." Interestingly, those who fall into this category are often more successful than others in making the explorations work. They may be more open to thinking of math in new ways and more comfortable with the idea of learning from their students. They *become* math people! This is not to say that knowledge of mathematical content is unimportant. The more deeply you understand the math you are teaching, the more effective you can be. However, as I suggested before, this learning occurs over time, and you do it best by listening to your students and reflecting on your practice. I can personally vouch for this. My own understanding of mathematics has been completely transformed by my work with elementary and middle school students.

I cannot promise that using these explorations will be easy at first. Nor can I promise that all of your students will love doing these types of activities right away. Some will thrive immediately. Many will adapt fairly quickly. But, understandably, a few will initially (and sometimes stubbornly) prefer what is familiar and comfortable.

What I can say is that, with a little faith and persistence, doing activities like these can change students' approach to math in profound and positive ways. They find themselves slowly drawn into this type of thinking and begin to miss it when it is not present! It can be a transformative experience. Students may leave your class with an entirely new understanding of mathematics as a discipline and of themselves as mathematicians. Personally, watching this transformation take place has been the greatest joy in my work. I wish the same joy for you.

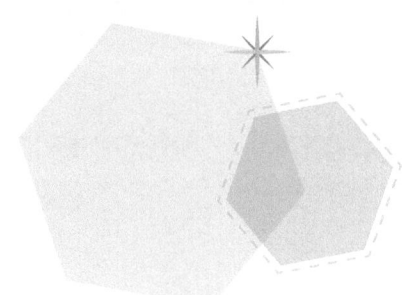

# Introduction

As educators, we often take one of two approaches to math with gifted students: a serious one, in which we stick to the curriculum and accelerate to the next topic (or the next course) or a fun one, with enriching games, puzzles, and general problem solving. Both approaches can benefit students when used appropriately. However, they leave out something important.

Acceleration alone can be superficial. It may not lead to deep, meaningful learning that lasts. Enrichment by itself may be unfocused. Students develop thinking skills, but the skills are disconnected from relevant mathematical content. With a careful blend of the two approaches, you get the advantages of each and something new—a deeper, richer understanding of advanced content that prepares students for long-term success and nurtures a love of math.

The explorations in this book are both serious and fun! In some of them, the fun part jumps right out at you. But the real fun starts after you spend time thinking about the problems and begin to unravel their secrets. Motivation spurs hard work. But, with the right tasks, hard work can also create motivation!

There is a simple teaching philosophy that captures the spirit of these explorations: the students do the thinking. When we tell talented students what to do, they do it, and it's done. When we let students think for themselves, they *learn* the math. They make sense of it. They retain it, transfer it to new learning, apply it to the real world, and appreciate it. As teachers, what more could we want?

# Connections to the Common Core State Standards

## THE COMMON CORE STATE STANDARDS FOR MATHEMATICAL CONTENT

Table 1 shows the Common Core State Standards for Mathematical Content that apply to each exploration. Because the problems are rich and open-ended, the standards for each activity naturally cross grade levels and content strands. This helps students make connections between math concepts. It also makes the explorations good vehicles for differentiation.

Try to be thoughtful in interpreting the Common Core grade-level designations. The authors of the CCSSM stated, "No set of grade-specific standards can fully reflect the great variety in abilities, needs, learning rates, and achievement levels of students in any given classroom" (NGA & CCSSO, 2010, p. 4). The learning trajectories of individuals will vary based on their experiences. Some talented students, especially those who have had many opportunities to learn through problem solving, may accelerate or even change the sequence of some learning trajectories (Johnsen, Ryser, & Assouline, 2014). Use your observations of students to make decisions about when to use an exploration or which parts of it to assign.

For example, because ratios and proportions appear prominently in grades 6 and 7 of the standards, many of the activities in this book are centered on these two grade levels. However, they may be appropriate for younger students who are more advanced. You can also differentiate the activities to make them meaningful for a broader age range. In Explorations 2 and 3, students are asked to explore ratios and rates from many points of view. Those who are newer to the concepts may focus on earlier standards—creating tables and diagrams, finding patterns in them, and connecting them to calculations. More advanced or capable students may spend more time connecting the tables to graphs and algebraic formulas, learning how to recognize proportional relationships in different contexts, and using ratios to explore the idea of slope.

## TABLE 1
Common Core State Standards for Mathematical Content

| Exploration | Standards |
|---|---|
| 1. The Incredible Shrinking Universe | **5.NBT.A, 6.RP.A.3, 7.G.A.1**<br>5.M.D.A, 5.M.D.B |
| 2. Ramps, Paints, and Hot Air Balloons | **6.RP.A, 6.E.E.C, 7.RP.A**<br>5.OA.B, 5.G.A, 8.EE.B |
| 3. Gear Up! | **6.RP.A, 7.RP.A**<br>5.OA.B, 5.G.A, 6.EE.A, 7.G.B.4 |
| 4. Perplexing Percentages | **6.RP.A, 6.EE.A, 6.EE.C, 7.RP.A**<br>7.EE.A.2 |
| 5. Scaling a Tower | **7.G.A**<br>6.EE.A, 8.EE.C.8 |
| 6. Keep It in Proportion | **6.EE.A, 7.RP.A**<br>5.NF.B |
| 7. Grab Bag | **6.RP.A, 6.EE.C, 7.RP.A, 7.EE.B**<br>6.EE.B |
| 8. Expanding and Contracting | **7.G.A, 8.NS.A**<br>6.G.A.1 |
| 9. Pythagorean Connections | **8.G.B, 8.EE.A.2, 8.EE.C.7**<br>7.G.A, 8.NS.A |
| 10. Twist and Shrink | **7.G.A, 8.G.A**<br>5.G.A, 6.G.A.3 |

*Note*: The standards shown in bold form the focus of the exploration.

# THE COMMON CORE STATE STANDARDS FOR MATHEMATICAL PRACTICE

The processes by which students learn and do mathematics are addressed in the eight Common Core State Standards for Mathematical Practice (NGA & CCSSO, 2010, pp. 6–8):

1. Make sense of problems and persevere in solving them.
2. Reason abstractly and quantitatively.
3. Construct viable arguments and critique the reasoning of others.
4. Model with mathematics.
5. Use appropriate tools strategically.
6. Attend to precision.
7. Look for and make use of structure.
8. Look for and express regularity in repeated reasoning.

Johnsen and Sheffield (2013) have proposed a ninth standard to support the development of mathematical innovators: "Solve problems in novel ways and pose new mathematical questions of interest to investigate" (pp. 15–16).

The practice standards, including Johnsen and Sheffield's proposed creativity standard, are deeply embedded in the explorations. Most of these standards are present in some form in every activity. However, bringing them to life in your classroom depends at least as much upon instruction as it does upon the activities themselves.

The first step is to remember that students must be the ones doing the thinking. The Teacher's Guide in the following section provides specific support for building the mathematical practices into your instruction: teaching and motivation strategies, ideas for developing mathematical communication skills, suggestions for managing classroom conversation, and examples of implementing the explorations in different settings. The explorations themselves contain lists of observations and questions to support probing conversation, samples of multiple thinking strategies, and some actual examples of classroom discussions. Finally, the assessment tool on page 18 is targeted specifically to objectives that will keep you and your students focused on the deep, concept-oriented learning that is at the heart of the Common Core State Standards for Mathematical Practice.

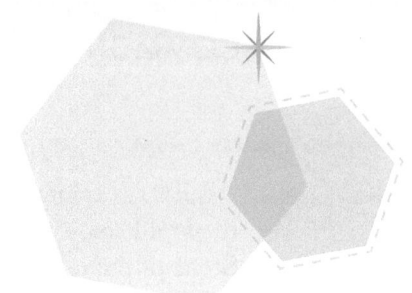

# Teacher's Guide

## GOALS

The explorations in this series were developed through years of work with talented middle school math students. They are designed to:

- » engage students in the excitement of mathematical discovery;
- » deepen students' understanding of middle school math concepts;
- » help students become flexible, creative, disciplined mathematical thinkers;
- » improve mathematical communication skills;
- » explore connections between math concepts;
- » develop patience, perseverance, and stamina in solving math problems;
- » provide depth and challenge for a variety of needs and interests;
- » enable students to work collaboratively and independently; and
- » offer opportunities for further exploration.

## THE EXPLORATIONS

This book contains problems that will challenge virtually any middle school math student. The explorations are self-differentiating. As students progress through an activity, the level of challenge or depth increases. A few students may finish. Most will reach a stopping point.

Students will progress according to their age, mathematical experience, persistence, capacity, and the amount of time available. Some may want to give up quickly. A few may insist on completing every problem even if they do not understand them well. A simple rule of thumb is that students should spend most of their time working on problems that are just beyond their comfort level. When they reach these problems, they should stick with them for a long time. They will learn more from thinking deeply about one or two problems than from rushing to finish a lot of them.

The problems in each exploration are grouped into three stages. Each successive stage extends the depth or the level of challenge. The end of a stage is a convenient place to pause and consider whether to continue. To help you decide, every stage starts with a brief description of the problems it contains along with information about the knowledge students will need and a summary of what they will learn.

Every exploration has a number of features to support your work with students: the Problems, some Conversation Starters, the Solutions, and Algebra Connections. Some of them have an additional feature called a Classroom Vignette. These features are described next.

# THE PROBLEM PAGE

Each Problem page has an "opener" and a list of directions. The opener is a sort of teaser that sets up the problem situation without telling students what to do. The directions fill in the details.

You may use the Problem page as a handout. My favorite approach is to cover the directions before I copy it so that students see only the opener on the handout. This is much more fun than giving them all of the information upfront. As we discuss the opener, students actively participate in creating the task by predicting (and suggesting) what the directions will be. This helps them learn that math is about asking questions, not just giving answers. It also allows me to identify possible points of confusion at the start.

At the end of this discussion, we finalize the directions. Based on students' ideas and their learning goals, we either modify the original directions or use them as is.

Some of the Problem pages have Testing the Waters or Diving Deeper questions at the bottom of the page. Testing the Waters questions are less complex versions of the main problem. They make it accessible to more students. If students are not making progress on the original problem, you can suggest that they begin with Testing the Waters. Even if they get no further, they will learn important new ideas.

The Diving Deeper questions are just what they sound like—an opportunity to explore in more depth. Many of these are more challenging than the original problem. Others point students to related questions or topics of interest.

# THE CONVERSATION STARTERS PAGE

The Conversation Starters are observations and questions that can or should arise in discussion. Sometimes, your students will come up with these. At other times, you will need to work them into the conversation. Their purpose is to help you guide your students' thinking without telling them how to do the problems.

You will probably not use all of the Conversation Starters on the page. Choose those that best fit your students' needs and learning goals, or follow up on the ones that your students initiate. You do not have to use them in any particular order, but I have tried to organize them in a way that is likely to follow the flow of discussion. The Conversation Starters near the end of the page are often extensions of the main ideas.

I have written the Conversation Starters as "I wonder" questions and "I notice" statements. In keeping with the philosophy of encouraging independent thinking, many of the "I wonder" questions are not answered. I have left them open for you and your students to think about. Even when you are not sure of the answer, the question may point your thinking in a useful direction. In some cases, especially with the items near the end of the page, I raise a question out of curiosity, and I may not know the answer myself.

By the way, "I wonder" questions may pop up in the Solutions, too! You are never done with a good problem. There are always more questions to ask!

# THE SOLUTIONS PAGE

In writing the Solutions, I have tried to strike a balance between giving enough detail to support your work and not giving so much that it makes the problems look harder than they are. Most solutions are one or two pages long. There are two main reasons for their length: (1) I include many student strategies, and (2) the problems contain a lot of ideas.

Please keep in mind that longer answers do not necessarily mean more classroom time. In some cases, I have simply shown many ways to think about the problem. On the other hand, a one- or two-line answer in the Solutions may represent a lot of thought and discussion. Although a solution may look short on paper, there is no shortcut for the effort and thinking that goes into finding it.

The Solutions are not the final word! You and your students may discover more efficient or more interesting strategies than I have shown. You will have insights that have not occurred to me. Each time I teach the explorations, I learn something new about the math.

# THE ALGEBRA CONNECTIONS PAGE

I have written most of these explorations assuming that students have a prealgebra level of knowledge—that they can understand, interpret, and even create algebraic expressions and equations, but they have learned few rules for manipulating them. Because your students will vary in their knowledge and experience, I have included an Algebra Connections page at the end of most activities. It has three purposes:

- » To help you see connections to students' future learning.
- » To give prealgebra students a chance to try their hand at algebraic processes and reasoning before they are taught all of the "steps" in algebra class!
- » To offer students who have studied algebra a chance to apply their skills to the problem.

If the Algebra Connections page does not seem relevant to your purposes, you may ignore it. You will not need it for other explorations. However, I hope you will glance at the connections between the content you are teaching and the concepts your students will study when you are no longer their teacher. You may gain valuable perspectives that inform your teaching. And if you feel comfortable doing so, allow your prealgebra students to play with some of the algebraic expressions and equations from time to time (without teaching them the rules)! This is a powerful way to integrate their understanding of numbers and variables.

A comment on notation: For the sake of consistency, I use a dot for multiplication and "÷" for division in most of the book. There are two exceptions. For scientific notation, I use "×" for multiplication, because it is traditional and easier to read. In the Algebra Connections, I often omit the dot and write division as a fraction. While teaching, I use a variety of notations in order to help students make the transition to traditional algebraic notation.

## THE CLASSROOM VIGNETTES

Some explorations have a sample conversation called a Classroom Vignette. Most of these are taken from actual conversations that I have had or witnessed with students. In some cases, I have combined ideas from conversations that took place over a couple of days. I have streamlined the flow of the discussions to make them more readable. And, of course, I have made up the names of the teachers and students. To make the conversations easier to write and read, there are usually just three or four students talking.

The main purpose of the vignettes is to give you a feel for how conversations might look when you are helping students to think independently and to develop conceptual understanding. I have chosen to write them on topics and problems that I felt could use a closer look, often because the Conversation Starters and the Solutions did not quite seem to capture certain key ideas.

I have not tried to make the vignettes illustrate every strategy for conducting effective discussion. I have worked harder to make them realistic than to make them perfect. My main hope is that these examples flesh out some important mathematical ideas for you and that they give you a starting point for thinking about ways to engage your students in conversations that help them make sense of math in a deep way.

## EIGHT MOTIVATION STRATEGIES

1.  **Let students know what to expect.** Tell students that the problem or activity will take time. Let them know that they will sometimes get stuck and that their work will probably not be perfect. Give them a time frame, and let them know how you will support them.

2. **Redefine success.** Tell students that success is not just about speed and accuracy. Let them know that you value effort, progress, creativity, insight, and clear communication—in short, you care more about learning than perfection.

3. **Praise effort over ability.** Praising effort over ability encourages risk-taking. Seeing intelligence as a quantity that changes through effort empowers students to reach their potential. Carol Dweck develops these ideas in her book *Mindset* (2007).

4. **Focus on process more than answers.** Respond to right and wrong answers in a similar manner, focusing on the mathematical ideas and the opportunity to learn something new. Show students that you value an interesting question as much as an accurate answer.

5. **Offer emotional support.** Some talented math students do not accept real challenges due to a fear of not looking "smart." They may not be accustomed to feeling frustration. They need help managing these feelings, especially if math has always come easily to them.

6. **Offer meaningful responses to written work.** You do not have to write a lot, just a few specific and thoughtful comments on students' completed work to let them know that you have read and considered their ideas.

7. **Allow students to collaborate.** In addition to the enjoyment of social interaction, collaboration makes students feel safer taking risks. And, of course, they have more success, because they are sharing ideas!

8. **Debrief.** After you finish a problem (or set of problems), talk about it before you move on. Kids love this! Share answers and strategies. Talk about what went right and what went wrong. Summarize key ideas. Discuss things that are still confusing. Think of new questions to ask.

## TEACHING STRATEGIES

Math is about ideas! Of course, skills are necessary, too, but without a conceptual foundation, students will not be able to apply skills to problems or use them to support further learning.

Shifting from a focus on procedural skills to a more balanced approach that recognizes the key role of ideas requires thinking in new ways. The strategies in the right hand column of Table 2 show how to use these explorations to support conceptual understanding and to infuse new depth and meaning into your students' learning.

## CLASSROOM DISCUSSIONS

The explorations are designed so that students may spend much of their time working without direct instruction. However, they will need to talk about the prob-

**TABLE 2**

Teaching Strategies

| Traditional Strategies | Strategies That Support Deep Learning |
|---|---|
| Prepare students for guided practice by clearly explaining procedures using worked examples. | Expect students to learn by thinking their way through challenging problems that engage them with the concepts. |
| Teach skills first. Then have students apply them to story problems. | Use problem solving as a means of teaching concepts and skills. |
| Grade homework by marking answers right or wrong. | Respond to students' work by writing comments related to their thinking. |
| Study answers in advance so that you can explain them clearly to the students. | Be ready to discuss unexpected strategies and learn new ideas from students. |
| Know the process you want students to use. | Assign tasks that can be solved in many ways. Discuss advantages and disadvantages of different methods. |
| Have every student do the same questions. | Differentiate goals and assignments based on students' learning needs. |
| Have fixed deadlines for assignments. | Be flexible with due dates if students run into unexpected difficulties or want to explore further. |

lems with you and with each other. You may need to be creative to find time for discussion, especially if you have a classroom with a wide range of needs, but it is worth the effort. The more that you and your students talk about the math, the more progress they will make and the more they will learn.

Equally important is what happens during conversation. Fortunately, you do not have to explain how to do the problems. That is your students' job! Yours is first to ensure that they have the basic knowledge needed to approach the problem and then to orchestrate conversation so that they learn from each other. The Conversation Starters give examples of questions and observations that move a discussion forward without telling students what to do. When in doubt, ask rather than answer, and say less rather than more.

To make conversations productive, classrooms must have a culture of curiosity and respect. All contributions to discussion are valuable, because all have the potential to create learning. Give students plenty of "wait time" before and after you call on them so that they have time to think and to formulate their responses. Ask them to speak in a strong voice and to direct their comments to the class rather than to you. Have them question, repeat, or rephrase each other's statements as needed. Have them agree or disagree—always explaining why. To facilitate, you may record and organize their ideas on the board. Rephrase their statements yourself for clarification if necessary, but always check that you have understood their ideas correctly. To learn more about these and other techniques for questioning and orchestrating classroom discussions, see Chapin, O'Connor, and Anderson (2013) and Smith and Stein (2011).

# ASSESSING STUDENT LEARNING

The tool on p. 18 is designed to assess concept-focused tasks. It was informed and inspired by many sources: the Common Core Standards for Mathematical Practice (NGA & CCSSO, 2010), the Process Standards of the National Council of Teachers of Mathematics (NCTM, 2000), the five Proficiency Strands in *Adding It Up* (Kilpatrick, 2001), and a rubric in *Extending the Challenge in Mathematics* (Sheffield, 2003).

You may design your own scoring system. I use a 5-point scale in each category.

5 evidence of learning beyond the level of course standards
4 evidence of learning at the level of course standards
3 evidence of learning approaching the level of course standards
2 evidence of learning below the level of course standards
1 evidence of learning significantly below the level of course standards
0 little or no evidence of progress toward meeting course standards

In my classes, students who are new to the explorations often receive 2s and 3s at first. As the school year progresses, they receive mainly 3s and 4s with an occasional 5. Students and parents appreciate the opportunity to identify specific areas of strength and goals for improvement. However, no numerical scoring system will ever replace the value of a few thoughtful written comments related to students' ideas!

Of course, you may also incorporate criteria such as legibility, organization, mechanics (spelling, punctuation, and grammar), etc. Above all, however, your scoring system should reflect the central goal of mathematical learning.

# MATHEMATICAL COMMUNICATION: PART 1

Learning to communicate mathematically offers two key benefits for students. It helps them to develop their own thinking and to communicate with others. This page focuses on the first reason. When I ask young elementary students why they believe it is important to explain their thinking, they usually mention both reasons. When I ask older students, they tend to focus on the second reason. I suspect that the more they learn to think of math only as numbers and calculations, the less value they place on thinking and writing.

You have probably experienced the challenge of trying to convince some of your students to write their ideas down. They may take pride in their ability to do it all in their heads at lightning speed. This type of intuition is wonderful, but it is not always reliable, and it is not enough. When students are working on problems that are sufficiently demanding to be worth their time, there is usually too much information to manage mentally. They must write as they work in order to remember what they have done, to clarify their thoughts, to visualize relationships, to recognize and extend patterns, and to identify and correct errors. I recommend that from the beginning students have paper and pencil in front of them at all times when they are solving challenging problems. In my classes, we call it "thinking paper" rather than scratch paper in order to emphasize its purpose and its importance.

A colleague of mine who is a high school English teacher tells her students that "fuzzy writing means fuzzy thinking." I cannot think of a truer statement for math class. The effort you take to write clearly helps you to think more clearly!

# MATHEMATICAL COMMUNICATION: PART 2

In order to express their ideas when solving deep and challenging problems, your students may need to expand their idea of what it means to "show work." Many important ideas cannot be captured in mathematical symbols alone. There are three common ways to communicate mathematical ideas: words, symbols (numbers, equations, etc.), and diagrams.

If students struggle with putting words on paper, suggest that they speak their ideas aloud and transfer them to paper. Of course, they will need to make some changes, but at least this gets them started. I often have students who insist that they do not know what to write—but are able to speak their thoughts perfectly clearly!

A few tips for using words:
» Use the word "it" sparingly, and always explain what "it" is!
» Can family or friends understand what you have written? If not, then rewrite.
» Be concise. More is not always better.

Students may be familiar with showing their thinking using numbers and equations, but there are still a couple of pitfalls. Labels or explanations of what numbers mean are important. Also, avoid "run-on" math sentences such as $13 - 7 = 6 \cdot 2 = 12$. This statement is false, and it shows a misunderstanding of the "$=$" symbol. It should be written $(13 - 7) \cdot 2 = 12$ or as two separate equations: $13 - 7 = 6$ and $6 \cdot 2 = 12$.

Diagrams can sometimes communicate ideas more clearly than words or symbols. They may even replace whole sentences or paragraphs! The key is to include all of the important information without cluttering them with unnecessary or distracting detail.

# EXAMPLES OF USING THE EXPLORATIONS

**Example 1:** Ms. Kava teaches gifted math pull-out groups of 5 to 10 students. Each group meets once per week during its regularly scheduled math time.

- » She coordinates with the classroom math teachers to select activities that align to course content.
- » Students work on explorations during the pull-out. They alternate between partner work and whole-group discussion.
- » Ms. Kava assigns independent "writing prompts" in which students summarize their understanding of a problem or solve a new problem from the exploration.
- » Ms. Kava shares her observations and students' completed work with the teachers.

**Example 2:** Mr. Hill teaches fifth-grade math in a cluster classroom with six identified high-ability math students. He has fewer students with other special needs, but he has a significant range of abilities in the classroom.

- » He uses flexible grouping. He holds two to three 15-minute math conversations with a small group of advanced students each week.
- » He uses pretests, exit slips, and informal observations to identify students for the advanced group. Cluster students participate regularly, while others flow in and out of the group as their needs indicate.
- » He makes the problems available to all who are interested. He sometimes discovers students with talents he had not noticed before.
- » Students work on the explorations in pairs on days he does not meet with them.
- » He has other enrichment tasks not requiring instruction for students to use when they are stuck on a problem and he is not available.
- » Most students work on Stage 1 of the explorations. More advanced or motivated students often continue further.
- » The school is planning to implement a half hour per day for targeted instruction when students are shared between classrooms. Mr. Hill and his colleagues plan to use the time for focused instructional opportunities with the explorations.

**Example 3:** Ms. Rodriguez teaches a stand-alone gifted math class for sixth graders who have scored at or above the 95th percentile on a standardized math test.

- » Much of the classroom instruction is based on the explorations.

» She uses her textbook primarily to sequence instruction and as a source of exercises to solidify concepts and skills.

» Many of the explorations are used as classroom lessons. Others are used as long-term (1- or 2-week) homework assignments.

» Lessons flow back and forth between small-group work on the problems and whole-group discussion in which students share and compare strategies.

» Ms. Rodriguez sets aside 10–15 minutes per day of time for class conversation about explorations that have been assigned as homework.

» Students typically complete Stages 1 and 2 of most explorations. She modifies expectations for a few students who are working hard but are having trouble finishing the homework. They complete fewer problems but are still expected to explain their thinking clearly and thoroughly. For students who need additional depth and challenge, Ms. Rodriguez makes Stage 3, Diving Deeper, or Algebra Connections tasks available as an option.

**Example 4:** Ms. Langford teaches the advanced section of a seventh-grade prealgebra class. Students in the class have generally scored at or above the 70th percentile on a standardized math test and had a supportive teacher recommendation. She implements a flipped classroom, in which students view daily instructional videos outside of class and complete summary tasks to verify understanding. Class time is used primarily to respond to individual needs.

» Based on her evaluation of the summary tasks, Ms. Langford creates groups whose members are in a similar place with respect to current learning goals.

» Some students spend a fair amount of group time on foundational understanding, but most are able to spend a significant portion of their time on the explorations.

» Ms. Langford takes about 10 or 15 minutes of class time to introduce a new stage of an exploration to the entire class when it comes up. Most students complete two stages of each exploration. A few complete all three. Sometimes, students skip Stage 1.

» As she monitors and assists with work on the explorations, Ms. Langford shifts students fluidly between groups in order to (1) allow those who are in similar places to work together in solving the problem, and (2) bring students into conversation when they have developed different ideas or strategies from which the other(s) can learn.

» Ms. Langford pulls groups together for a larger, focused conversation when she hears insights or misconceptions that many will benefit from discussing.

» She assigns one or two selected problems from the explorations to be written up, turned in, and graded each week.

**Example 5:** Mr. Okoro is the primary person responsible for teaching math at a small school. With one math class per grade level, he has a full range of needs in each class.

» He pretests students and offers the explorations to students who have demonstrated an understanding of the concepts from a lesson or unit.

» He reduces the number of practice exercises for these students and has them work in small groups on the explorations during their extra time in class.

» He meets with them as often as he can, but because he has limited time, he also makes "hint cards" from the Conversation Starters. If students have worked hard but are stuck, and he is not available to help, they take a hint card and use it to continue discussion with their peers.

» He sometimes introduces the first problem of an exploration to the whole class, using the Testing the Waters problem or other modifications to make it accessible to all. Occasionally, he discovers students who may not test well, but show signs of talent in solving nonroutine problems.

» The school has set aside a time early in the day when students meet across grade levels. He proposes using some of this time for students to receive targeted instruction and conversation time with the explorations.

# Assessing Student Learning

| Criterion | Description | Score |
|---|---|---|
| **Depth of Understanding** | » Know the *why* behind the *how*.<br>» Understand the meanings of concepts.<br>» Recognize and use connections between ideas. | |
| **Problem Solving** | » Create and use effective problem-solving strategies.<br>» Verify your results.<br>» Solve the problem more than one way. | |
| **Elaboration and Communication** | » Give thorough, clear, concise explanations.<br>» Use words, calculations, and diagrams effectively.<br>» Support your explanations with examples. | |
| **Generalizations and Reasoning** | » Recognize, analyze, and extend patterns.<br>» Make and test predictions.<br>» Use logic to evaluate claims and justify conclusions. | |
| **Correctness and Precision** | » Give correct answers stated with appropriate precision.<br>» Calculate accurately and efficiently.<br>» Use mathematical vocabulary correctly and precisely. | |
| **Originality and Extensions** | » Invent ideas and strategies that were not taught.<br>» Find ideas and strategies that are rarely discovered.<br>» Propose new ideas or questions to study. | |
| **Effort and Perseverance** | » Show consistent effort.<br>» Make progress appropriate to your understanding.<br>» Persist through difficulties. | |

# Exploration  1

## The Incredible Shrinking Universe

As space enthusiasts know, the universe is not really shrinking; it is expanding! But, in this exploration, students discover that they can use math to bring astronomical measurements down to a human scale and to imagine the incomprehensible!

The large numbers in these problems are the most interesting but also the most challenging feature for my students. I generally allow them to use calculators. At the same time, I encourage them to use their knowledge of place value to simplify the calculations, and I insist that they estimate before they calculate—to think about what sort of an answer they expect before they "crunch the numbers."

I placed this exploration at the beginning of the book because my students are able to enjoy and understand it before they have studied proportional relationships in depth. Talented younger students often succeed using their intuition and their knowledge of map scales. The activity provides them with an excellent set of experiences to support their emerging understanding of deeper concepts surrounding proportionality. If your students struggle early with the activity, return to it when they have had a little more experience with ratios or large numbers.

Many students have studied our solar system and are curious to learn more, but the sizes and distances involved are hard to imagine! Stage 1 will help them understand how the sizes of the planets compare.

Some students may not feel comfortable choosing their own scale and developing their own strategies. They may ask what the "best" scale and the "best" strategy are. It takes time and patience for students to learn that there is not always a best answer—and that even when there is, they are capable of discovering it for themselves! These are exactly the skills they need to learn in order to apply math to real-world situations.

## What Students Should Know

>> Understand multiplication and division of decimals.
>> Translate among customary and metric units of measurement.
>> Have experience working with large numbers.
>> Read and interpret a scale on a map (recommended).

## What Students Will Learn

>> Choose an appropriate scale for a scale model.
>> Use multiplication and division to create a scale model.
>> Apply knowledge of place value to calculate with large numbers.
>> Use intuition about scale models to further develop proportional reasoning skills.

# Problem #1

Makayla is creating a presentation to help her class-mates understand how the Earth compares in size to other objects in our solar system. She decides that using a scale model will make it easier for her classmates to visualize the comparisons.

## *Directions*

- Gather some data, and choose a scaled size for the Earth.
- Calculate the scaled sizes of the sun, planets, and some other objects in our solar system. Describe your thinking processes, and show your calculations.
- Offer Makayla some advice on presenting her ideas effectively.

### *Diving Deeper*

- If you used the radius or diameter to make your comparisons, try using circumference, surface area, or volume instead. You may be surprised at the results!

- Think of a situation when you could *enlarge* objects to make them easier to visualize and compare. Do the calculations, and describe what you have learned.

# CONVERSATION STARTERS FOR #1

*What do you notice? What do you wonder?*

*I notice* that the planets close to the sun are smaller than the more distant planets.

*I wonder* how I should decide on a scaled size for the Earth?

Consider making it the size of a familiar object. Then estimate what the other planets will look like on this scale.

*I notice* that I can mix units when I create a scale.

For example, 1 centimeter in my scale model can represent some number of kilometers in real life.

*I wonder* which objects I should include besides the planets?

Consider including the sun, the moon, dwarf planets, or asteroids.

*I wonder* how many decimal places I should show in my answers?

One possibility is to include just enough places to enable you to make a good drawing or physical model. Be sure to take into account the number of decimal places in the data you collected.

# SOLUTIONS FOR #1

Suppose the Earth were the size of a baseball, about 7.6 cm (0.076 m) in diameter. The actual and scaled diameters of the Sun and planets (and a few other objects) would be:

| Sun | 1,391,684 km | 829 cm (8.29 m) |
|---|---|---|
| Mercury | 4,879 km | 2.9 cm |
| Venus | 12,104 km | 7.2 cm |
| Earth | 12,756 km | 7.6 cm |
| Moon | 3,475 km | 2.1 cm |
| Mars | 6,792 km | 4.0 cm |
| Ceres (dwarf planet) | 950 km | 0.57 cm |
| Jupiter | 142,984 km | 85.2 cm |
| Saturn | 120,536 km | 71.8 cm |
| Uranus | 51,118 km | 30.5 cm |
| Neptune | 49,528 km | 29.5 cm |
| Pluto (dwarf planet) | 2,390 km | 1.42 cm |

Of course, other choices for a scale are possible, and this will affect the answers.

*Strategy 1*: Create a scale comparing real lengths in kilometers to scaled lengths in centimeters. Every centimeter in the scale model stands for 1678 km in the real world, because $12,756 \div 7.6 \approx 1678$. Therefore, Mercury's scaled size is approximately:

$$4879 \div 1678 \approx 2.9 \text{ cm}$$

*Strategy 2 (a variation of Strategy 1)*: Find a *scale factor* between the Earth and a baseball. The diameter of the Earth is approximately 12,756 km (12,756,000 meters) and the diameter of a baseball is about 7.6 cm (0.076 meters). The scale factor is:

$$12,756,000 \div 0.076 \approx 167,800,000$$

This means that the Earth's diameter is about 167,800,000 times the diameter of a baseball! Divide the diameters of the other planets by this scale factor. For example, Mercury's diameter is about 4,879,000 meters, so its scaled size is:

$$4,879,000 \div 167,800,000 \approx 0.029 \text{ m (or 2.9 cm)}$$

*Strategy 3*: Use division to compare the diameters of two planets. For example, to compare the Earth to Mercury (kilometers), you would use:

$$12,756 \div 4879 \approx 2.614$$

This means that the Earth's diameter is about 2.614 times Mercury's diameter. Divide the diameter of a baseball (in cm) by this number to find Mercury's scaled size:

$$7.6 \div 2.614 \approx 2.9 \text{ cm}$$

*Advice for Makayla's presentation*: To make the numbers easier to visualize, Makayla could compare the scaled sizes to familiar objects. For example, Mars would be about the size of a ping-pong ball; the sun would just about fill a typical classroom! Makayla could also create physical models of some of the planets to show to her classmates.

# STAGE 2

A realistic scale model for our solar system would use the same scale for distance as it does for size. Your students may discover that building such a model is not very practical, because it would be too large! It may make more sense to use a smaller scale for the distance. On the other hand, the size-based scale may work if they prefer simply to visualize the situation. The Solutions for Problem #2 show an example of each option.

In the Solutions, I show customary units for the sake of variety and familiarity. Students may choose either customary or metric units. If different students use different units, ask them to discuss the advantages and disadvantages of each.

As your students begin working, mention that the distances between the sun and the planets are averages. Because planetary orbits are not quite circular (they are *ellipses*), the distances vary. Some students may be interested in learning more about planetary orbits!

## What Students Should Know

> » Understand the concepts and strategies from Problem #1.

## What Students Will Learn

> » Continue to use scale models to develop proportional reasoning skills.
> » Compare the effects of using different scales.
> » Identify needed information, and solve complex multistep problems.

# Problem #2

Now that her classmates have a better understanding of the sizes of the planets, Makayla wants to help them imagine the distances between the planets.

## Directions

- Decide if Makayla should use the same scale as in the previous question. Explain your thinking.
- Choose a scale, and use it to calculate scaled distances between the sun and other objects in the solar system. Describe your thinking processes and show your calculations.
- Decide if it is practical for Makayla to create a physical model of the solar system to display in her school. If not, explain why. If so, explain how!

## Diving Deeper

- Read about *astronomical units*. How do they relate to this problem?

- Choose a planet and a spacecraft that has visited it. Estimate or calculate the travel time from the Earth. Compare your result to the actual time the spacecraft took to reach the planet. The answers may be very different! Why?

# CONVERSATION STARTERS FOR #2

*What do you notice? What do you wonder?*

*I notice* that the distances between planets increase as you move farther from the sun.

*I wonder* which objects I should include besides the planets?

Consider including an object from the asteroid belt and perhaps a distant dwarf planet such as Pluto. It may also be interesting to include a comet, but their unusual orbits may make this challenging!

*I wonder* whether I should use customary or metric units?

Customary units such as miles, feet, and inches may be more familiar. However, metric units may be easier to calculate with. For example, it is easier to see that 0.5334 m = 53.34 cm than it is to show that 2.57 feet is approximately 2 feet 7 inches.

*I notice* that the scaled distances are very large when I use the same scale that I used in Problem #1 for the sizes of the planets!

*I wonder* how to enter large numbers in my calculator if they don't fit on the display?

Be creative! For example, you could enter the number 2,799,000,000 as $2799 \cdot 1,000,000$. Or, if you know scientific notation, use it!

*I wonder* if there is a pattern to the distances between the planets?

In most cases, each planet seems to be a little less than twice as far from the sun as its inner neighbor. The distance between Mars and Jupiter is a striking exception! Including the dwarf planet Ceres in the list seems to "fill the gap." Some students may be interested in reading more about the asteroid belt.

# SOLUTIONS FOR #2

*Makayla's choice of scale*: If Makayla uses the same scale as she did for the sizes of the planets, the Earth will be more than a half-mile from the sun! Of course, it would not be practical to build this, but it might still help her classmates imagine the distances. To build an actual model, she would need to choose a smaller scale such as the length of a school hallway.

*Strategy 1 (using the same scale for both size and distance)*: Divide the actual distance by 167,800,000 (the scale factor from Strategy 2 in the Problem #1). For Mercury, this would be:

$$36,000,000 \div 167,800,000 \approx 0.21 \text{ miles}$$

The actual and scaled distances of the planets from the sun are:

| | | |
|---|---|---|
| Sun* | 0 miles | 0 miles |
| Mercury | 36,000,000 miles | 0.21 miles |
| Venus | 67,000,000 miles | 0.40 miles |
| Earth | 93,000,000 miles | 0.55 miles |
| Mars | 142,000,000 miles | 0.85 miles |
| Ceres (dwarf planet) | 257,000,000 miles | 1.53 miles |
| Jupiter | 484,000,000 miles | 2.88 miles |
| Saturn | 888,000,000 miles | 5.29 miles |
| Uranus | 1,784,000,000 miles | 10.63 miles |
| Neptune | 2,799,000,000 miles | 16.68 miles |
| Pluto (dwarf planet) | 4,670,000,000 miles | 27.83 miles |

*Students will discover in later explorations that *proportional relationships* such as this always begin with both values equal to 0!

*Strategy 2 (using a smaller distance scale)*: Set the distance between the sun and Neptune as the length of a school hallway—say about 200 feet. Every foot represents $2,799,000,000 \div 200 \approx 13,995,000$ miles. So Mercury's scaled distance is $36,000,000 \div 13,995,000 \approx 2.57$ ft. (Students may avoid changing decimals into feet and inches by using the metric system instead!)

| | | |
|---|---|---|
| Mercury | 36,000,000 miles | 2 feet 7 inches |
| Venus | 67,000,000 miles | 4 feet 9 inches |
| Earth | 93,000,000 miles | 6 feet 8 inches |
| Mars | 142,000,000 miles | 10 feet 2 inches |
| Ceres (dwarf planet) | 257,000,000 miles | 18 feet 4 inches |

| Jupiter | 484,000,000 miles | 34 feet 7 inches |
|---------|-------------------|------------------|
| Saturn | 888,000,000 miles | 63 feet 5 inches |
| Uranus | 1,784,000,000 miles | 127 feet 6 inches |
| Neptune | 2,799,000,000 miles | 200 feet |
| Pluto | 3,670,000,000 miles | 262 feet 3 inches |

*Strategy 3 (using the same distance scale as in Strategy 2)*: Mercury is about $36,000,000 \div 2,799,000,000 = 36 \div 2799 \approx 0.01286$ times as far from the sun as Neptune is. $200 \cdot 0.01286 \approx 2.57$ feet $\approx 2$ feet 7 inches.

# Problem #3

Proxima Centauri, the nearest star to our solar system, is approximately 4.3 *light years* from us. A light year is the distance that light travels in one year.

## Directions

- Use your distance scale (Problem #2) to find the scaled distance to Proxima Centauri.
- Calculate the time needed for a spacecraft to reach Proxima Centauri.

### *Diving Deeper*

For students who finished Stage 1 of Exploration 9: Multiplication Slide Rules in the *Numbers and Operations* book: Find the multiplication scale measurements for the real-world and scaled distances you calculated in this activity. What do you notice? What do you wonder?

# CONVERSATION STARTERS FOR #3

*What do you notice? What do you wonder?*

*I notice* that I need more information to answer the questions.

Yes! What do you need to know? (Perhaps the speed of light and typical speeds for planetary exploration spacecraft. Do some research!)

*I wonder* whether I should use customary or metric units?

Familiar units may help you visualize the situation better, but metric units may be easier to calculate with. It's your choice!

*I notice* that when numbers get too large, my calculator displays them in a new way.

Try to find a way to keep track of the place value yourself!

*I wonder* if it matters whether I calculate the travel time using real speeds and distances or scaled speeds and distances?

Try it both ways! If you use scaled distances, be sure to adjust the spacecraft's speed to fit the scale.

# SOLUTIONS FOR #3

*Proxima Centauri's approximate distance from our solar system:*

| Actual | 40,700,000,000,000 km | 25,277,000,000,000 mi |
|---|---:|---:|
| "Hallway" scale | 550 km | 342 mi |
| "Size" scale | 240,000 km | 150,000 mi |

*Time to reach Proxima Centauri at 57,600 km per hr (speed of Voyager 1 spacecraft):* About 81,000 years!

*A calculation process for the travel time:* Light travels approximately 300,000,000 meters in one second:

300,000,000 m per sec ÷ 1000 m per km = 300,000 km per sec

300,000 km per sec · 3600 sec per hr = 1,080,000,000 km per hr

1,080,000,000 km per hr · 24 hr per day = 25,920,000,000 km per day

25,920,000,000 km per day · 365.25 days per yr = 9,467,280,000,000 km per yr

9,467,280,000,000 km per yr · 4.3 yr = 40,709,304,000,000 km (distance to Proxima Centauri)

40,709,304,000,000 km ÷ 57,600 km per hr = 706,758,750 hr

706,758,750 hr ÷ 24 hr per day ≈ 29,448,300 days

29,448,300 days ÷ 365.25 days per yr ≈ 81,000 yr

Some students will find more efficient approaches. Doing a lot of rounding early in the process will affect the accuracy of their answers. Encourage them to keep more precision in their calculations than they plan to show at the end.

# CLASSROOM VIGNETTE FOR #3

Mr. Hill notices that his students are struggling with the calculations when the numbers get large. He decides to use this as an opportunity to talk about place value.

**Mr. Hill:** I noticed some people having trouble multiplying large numbers. For example, when they tried this calculation, their calculator showed a funny looking answer. (He writes $1,080,000,000 \cdot 24$ and the calculator display $2.592^{10}$ on the board.)

**Patrick:** Yeah, what does the little 10 on the right side mean?

**Mr. Hill:** We'll talk about that soon! For now, why do you think the calculator isn't showing the number the usual way?

**Jasmine:** Maybe because it's too big? It won't fit on the screen anymore.

**Mr. Hill:** Just for now, I wonder if we can figure out a way to do it without a calculator! Does anyone have an idea?

**Patrick:** We could take off the zeros and put them back when we're done. *(Other students nod.)*

**Mr. Hill:** It looks like some of you have done this before. How would you do it?

**Jamal:** Well, I'd take all the 0s off the end of 1,080,000,000 and turn it into 108. I'd multiply that by 24 and then put the 0s back on to the answer.

*(Mr. Hill records Jamal's ideas as he explains them, showing $108 \cdot 24 = 2592$ and then the final answer of 25,920,000,000.)*

**Mr. Hill:** Does everyone agree with that? *(The students nod.)* Let's think about this for a minute. What happened to the number 1,080,000,000 when we took away the 0s at the end? *(Students look unsure.)* What happened to its size?

**Jamal:** It got a lot smaller!

**Mr. Hill:** Can anyone add to that? *(He waits, but no one responds.)* Let's try it with some smaller numbers first. *(He writes on the board:  3000   300.)* What happens to the 3 when you remove the 0?

**Patrick:** It moves to the right.

**Mr. Hill:** What happens to its value? *(No one responds.)* Think about the place value.

**Patrick:** It goes from the thousands to the hundreds.

**Mr. Hill:** How does that affect the size of 3000?

**Katelyn:** It makes it get 10 times smaller.

**Mr. Hill:** Why does that happen?

**Jasmine:** Because now you have 3 hundreds instead of 3 thousands, and there are 10 hundreds in every thousand.

**Mr. Hill:** What happens if you remove two 0s from 3000?

**Jamal:** The 3 goes from the thousands to the tens. *(Mr. Hill waits to see if he will say more.)* That makes the 3000 get a hundred times smaller.

**Mr. Hill:** Why a hundred?

**Katelyn:** Because there are a hundred tens in every thousand.

**Mr. Hill:** Let's keep this going. What happens when we keep moving the 3 to the right? 3 places? 4 places? 5 places?

*(The class answers and Mr. Hill records and writes their responses on the board.)*

**Mr. Hill:** *(Summarizing and extending.)* As Patrick said, the 3 keeps moving one place to the right. Once it gets into place values less than the ones, we aren't removing 0s any more, but the value still keeps getting one tenth as large every time we move it one place! Now I think you can tell me what happens to the size of 1,080,000,000 after the 0s are removed!

*(The class returns to this discussion, noticing that the digit 1 shifts seven places from the billions to the hundreds place, and that each shift makes the number one tenth as large. Putting everything together, the number gets 1 ten millionth as large, which is like dividing it by ten million. They return to Jamal's strategy for multiplying 1,080,000,000 by 24. They rephrase it as dividing by 10 million, multiplying by 24, and then multiplying by 10 million.)*

**Mr. Hill:** I like the way we're thinking about place value and the sizes of the numbers. I think we're ready to look at the way that Patrick's calculator showed the answer. *(He writes it on the board next to the answer that they calculated earlier.)*

Calculator display: $2.592^{10}$     Our answer: 25,920,000,000

**Mr. Hill:** What is the value of the 2 in 2.592?

**Patrick:** It's just 2 because it's in the ones place!

**Mr. Hill:** What is the value of the 2 in 25,920,000,000?

**Jasmine:** 20 billion, because it's in the ten billions place.

**Mr. Hill:** How many places did the 2 move from the ones to the ten billions? *(Students count the places.)*

**Patrick:** It moved 10 places. Oh, I see! That's what the small 10 on my calculator means!

*(Mr. Hill summarizes the discussion and tells the students that the calculator is showing the number in* scientific notation, *something they will learn more about later. He points out that calculators display scientific notation in different ways. He encourages his students to apply what they learned from the discussion to develop strategies for calculating with the large numbers, with or without a calculator. He also challenges them to think about place value instead of removing and putting on zeros. He knows that they will still*

*have some struggles, but he wants to observe their thinking as they apply what they have learned so far.)*

**Mr. Hill:** I have one more thing for you to think about before we go back to the problem about Proxima Centauri. What happens when you remove 0s from the numeral 0.3000? Why is this different than what happens when you remove them from 3000?

# STAGE 3

In the 1980 television series, *Cosmos*, astronomer Carl Sagan introduced viewers to the "Cosmic Calendar" in which we imagine that the history of the universe takes place within a single year! In Stage 3, students carry out Cosmic Calendar calculations using current data presented by Neil deGrasse Tyson in a 2014 production of *Cosmos*. The results illustrate our place in history on a cosmic scale!

The Cosmic Calendar problem offers a few new challenges: (1) the data contain a greater range of numbers, (2) some of the time conversions require more thought, and (3) students must be careful when they interpret the meanings of their results. For example, 3.7 days from the beginning of the Cosmic Calendar occurs on January 4, not January 3.

Students who are familiar with scientific notation may solve this problem more easily, but all students will have an opportunity to think deeply about division and place value! Ideas for using place value to develop division strategies are discussed in the Conversation Starters.

## What Students Should Know

» Understand the concepts and strategies from Problems #1–#3.
» Knowledge of scientific notation may be helpful but is not necessary.

## What Students Will Learn

» Apply knowledge of scale models to a new situation with a larger range of numbers.
» Use place value concepts to develop strategies for large number calculations.

# Problem #4

In the 1980 television series, *Cosmos*, astronomer Carl Sagan introduced viewers to the Cosmic Calendar, which collapses the entire history of the universe into a single year beginning at midnight on January 1! The data below are selected from the more recent *Cosmos* series featuring astronomer Neil deGrasse Tyson.

| Event | Years Ago | Cosmic Calendar |
|---|---|---|
| Beginning of the universe | 13.8 billion | January 1, 12:00 a.m. |
| Formation of the Milky Way galaxy | 11 billion | |
| Formation of our sun | 4.57 billion | |
| First life | 3.8 billion | |
| The first fish | 500 million | |
| The first reptiles | 300 million | |
| Cretaceous extinction of dinosaurs | 65 million | |
| First use of stone tools by humans | 2.5 million | |
| Beginning of the most recent ice age | 110 thousand | |
| Agriculture | 12 thousand | |

*Directions*
- - - - - - - - - -
- Determine the date on which each event occurred on the Cosmic Calendar.
- Calculate the length of a typical human life on this scale.

# CONVERSATION STARTERS FOR #4

*What do you notice? What do you wonder?*

Many of the Conversation Starters from Stages 1 and 2 apply here as well. This page focuses on division concepts that help when numbers are too large to fit on your calculator's display. Students may have already seen these ideas when they learned to divide decimals. They may even have applied them in earlier problems, whether they realized it or not! A quick vocabulary review: *dividend ÷ divisor = quotient*.

*I wonder* how changing the dividend affects the quotient?

When you increase the dividend, the quotient becomes larger. For example, if the dividend becomes 10 times larger, the quotient does the same.

*I wonder* how changing the divisor affects the quotient?

When you increase the divisor, the quotient become smaller. For example, if the divisor becomes 10 times larger, the quotient becomes one tenth as large.

*I wonder* how changing the dividend and the divisor affects the quotient?

When you multiply (or divide) the dividend and the divisor by the same number, the quotient remains the same, because one change compensates for the other.

*I notice* that I can apply these ideas to large number calculations.

For example, to divide 10 billion by 13.8 billion, you can just divide 10 by 13.8, because the dividend and divisor have both been divided by the same number, 1 billion.

*I wonder* what happens if I multiply the dividend and divisor by different numbers?

The quotient will change—but in a predictable way. For example, if the dividend becomes 1000 times larger, but the divisor become only 100 times larger, the net effect is to make the quotient 10 times larger. (Think carefully to understand why!)

# SOLUTIONS FOR #4

| Event | Years Ago | Cosmic Calendar |
|---|---|---|
| Beginning of the universe | 13.8 billion | January 1, 12:00 a.m. |
| Formation of Milky Way galaxy | 11 billion | March 16 |
| Formation of our sun | 4.57 billion | September 2 |
| First life | 3.8 billion | September 22 |
| The first fish | 500 million | December 19 |
| The first reptiles | 300 million | December 24 |
| Cretaceous extinction of dinosaurs | 65 million | December 30, midday |
| First use of stone tools by humans | 2.5 million | December 31, 10:25 p.m. |
| Beginning of most recent ice age | 110 thousand | December 31, 11:55:49 p.m. |
| Agriculture | 12 thousand | December 31, 11:59:33 p.m. |

A human life (80 years) lasts less than a fifth of a second in Cosmic Calendar time! (Some answers may vary slightly due to rounding choices.)

*Strategy 1 (for the beginning of agriculture):*

| Cosmic Calendar Time | Real Time | Calculation |
|---|---|---|
| 1 day | 37,800,000 years | 13.8 billion ÷ 365.25 |
| 1 hour | 1,575,000 years | 37,800,000 ÷ 24 |
| 1 minute | 26,250 years | 1,575,000 ÷ 60 |
| 1 second | 438 years | 26,250 ÷ 60 |

The beginning of agriculture occurs $12,000 \div 438 \approx 27$ seconds before the end of the Cosmic Calendar year, which is 11:59:33 pm on December 31.

*Strategy 2 (for the first appearance of life):* Because life appeared about 3.8 billion years ago, it first occurred around 10 billion years into the history of the universe. The fraction of total lifetime of the universe to that point is about:

$$10 \text{ billion} \div 13.8 \text{ billion} = 10 \div 13.8 \approx 0.7246$$

This represents $0.7246 \cdot 365 \approx 264.5$ days. Thus, 264 days have passed, and we are about halfway into the 265th day of the year. In a non-leap year, the 265th day is September 22.

# Exploration 2

## Ramps, Paints, and Hot Air Balloons

I like doing this exploration with kids, because they get to draw pictures to make and test their predictions about complex ideas. In the first stage, they use ramps to begin investigating the concept of slope while extending their understanding of ratios. Because their pictures look a lot like graphs, they start to understand how the steepness of a graph relates to patterns of change they see in the numbers.

Ratios, rates, and proportions offer wonderfully rich opportunities for students to develop and discuss thinking strategies. This exploration contains a Classroom Vignette in which students explore one of the questions in the Conversation Starters for Problem #2. Notice how, with careful questioning and support, students are able to develop their own procedures for naming steepness with a number. Encourage them to create and compare their own strategies for all of the problems! With your guidance, they will gradually discover effective and efficient methods that make sense to them.

# STAGE 1

In the three Stage 1 problems, students use tables, pictures, graphs, and calculations to make sense of situations involving ratios and rates in real-world contexts. I begin by asking them to imagine what Lupe's ramps could be used for. They often suggest things like bikes, skateboards, toy cars, roller coasters, grain elevators, etc. It is fun for them to picture these as they work through the problems!

Before you begin Problem #3, be sure that students know the ":" symbol for *ratios*. For example, 2 boys : 4 girls means 2 boys *for every* 4 girls. Ask students to name other ratios equivalent to this. (Possible answers: 1 boy : 2 girls or 5 boys : 10 girls)

## *What You Will Need*

»   Plenty of graph paper.

## *What Students Should Know*

»   Multiply and divide numbers when the answer is a fraction or decimal.
»   Write basic fractions as percentages.
»   Plot points on a coordinate grid.

## *What Students Will Learn*

»   Recognize situations when ratios apply.
»   Represent ratios with pictures, tables, and graphs.
»   Solve challenging problems involving ratios.
»   Identify the *rate* associated with equivalent ratios.
»   Connect ratios and rates to the concept of steepness or *slope*.

# Problem #1

Lupe's company designs and builds multipurpose ramps of all sizes. One of her ramps looks like this.

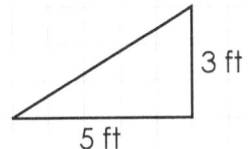

5 by 3 ramp

## Directions

- Show two larger and two smaller ramps that have the same steepness as this ramp.
- Explain your answers with pictures or graphs.
- Make a table showing ramps with this steepness. Describe any patterns you see.
- Explain your answers with calculations.

    Suggestion: Use *H* and *V* for the horizontal and vertical side lengths, respectively.

## Testing the Waters

Solve Problem #1 for a 2 by 1 ramp.

*Advanced Common Core Math Explorations: Ratios, Proportions, & Similarity* © Taylor & Francis.

# CONVERSATION STARTERS FOR #1

*What do you notice? What do you wonder?*

*I notice* that the ramps are named as "*H* by *V*" (with the horizontal length listed first).

*I wonder* if it matters whether I use feet, inches, or some other unit?

*I wonder* what happens to the steepness when I add the same number to *H* and *V*?
  Try changing a 5 by 3 ramp into a 6 by 4 ramp—or into a 105 by 103 ramp!

*I notice* that when I multiply *H* and *V* by the same number I get a ramp with the same steepness.

*I notice* that *H* and *V* make equivalent fractions in my table!

*I notice* a pattern in $H - V$ for my ramps!
  $H - V$ equals 2 for a 5 by 3 ramp, 4 for a 10 by 6 ramp, and 6 for a 15 by 9 ramp, etc.

*I wonder* what causes this pattern?

*I notice* that my graph looks like my picture of the ramp!

*I notice* that when *H* equals 1 foot, *V* equals $\frac{3}{5}$ of a foot.

*I wonder* if it is a coincidence that the numerator of $\frac{3}{5}$ is *V* and the denominator is *H*?

*I notice* that *V* is always $\frac{3}{5}$ of *H*, and *H* is always $1\frac{2}{3}$ times *V*.

*I notice* that $\frac{3}{5}$ and $1\frac{2}{3}$ are reciprocals!

# SOLUTIONS FOR #1

*Possible ramp sizes*: 10 by 6, 15 by 9, $2\frac{1}{2}$ by $1\frac{1}{2}$, 1 by $\frac{3}{5}$

*A picture/graph*:

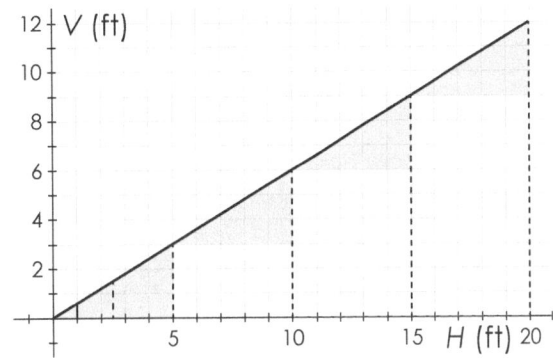

The light grey triangles show how students often use copies of the original ramp to guide the construction of larger ones.

*Some tables and patterns*:

| H (ft.) | 5 | 10 | 15 | 20 | 25 | 30 |
|---------|---|----|----|----|----|----|
| V (ft.) | 3 | 6 | 9 | 12 | 15 | 18 |

| H (ft.) | 1 | 2 | 3 | 4 | 5 | 6 |
|---------|---|---|---|---|---|---|
| V (ft.) | 0.6 | 1.2 | 1.8 | 2.4 | 3.0 | 3.6 |

» The numbers in each row increase by the same amount each time.
» The value of *V* when *H* equals 1 foot is helpful for finding other values of *V* (0.6 or $\frac{3}{5}$ is the *rate* for this situation).
» The horizontal side length times 0.6 always equals the vertical side length.

*Some calculations*:

» Multiply or divide 5 and 3 by the same number.

$$5 \cdot 2 = 10 \text{ and } 3 \cdot 2 = 6 \text{ (10 by 6)} \qquad 5 \div 5 = 1 \text{ and } 3 \div 5 = \frac{3}{5} \text{ (1 by } \frac{3}{5}\text{)}$$

» Divide a horizontal side by 5, and multiply by 3 to get the vertical length.

$$10 \div 5 \cdot 3 = 2 \cdot 3 = 6 \text{ (10 by 6)} \qquad 1 \div 5 \cdot 3 = \frac{1}{5} \cdot 3 = \frac{3}{5} \text{ (1 by } \frac{3}{5}\text{)}$$

# Problem #2

These three ramps are Lupe's top sellers.

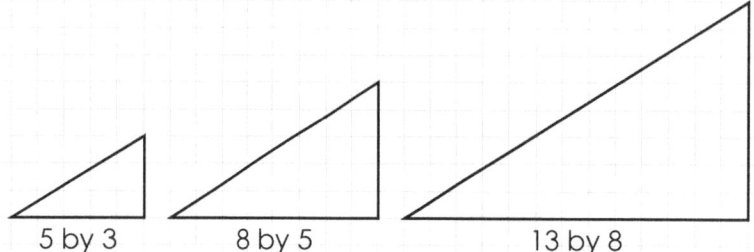

5 by 3          8 by 5          13 by 8

## Directions

- Decide if any or all of the ramps have the same steepness. If not, arrange them from flattest to steepest.

Do the following things in any order:
- Support your answers using tables.
- Support your answers using pictures or graphs.
- Support your answers using calculations.

## Diving Deeper

How do your answers to Problem #2 change if Lupe tips the ramps to make the horizontal sides vertical?

## Testing the Waters

Arrange these ramps from flattest to steepest: 2 by 1, 3 by 2, 5 by 3.

# CONVERSATION STARTERS FOR #2

*What do you notice? What do you wonder?*

*I notice* that the steepness of all three ramps looks about the same.

> *Suggestion*: Do not rely only on the pictures on the problem page. Make your own drawings on graph paper, use a sharp pencil, and look very closely!

*I notice* that ramps are easier to compare when one pair of sides is the same length.

*I wonder* if a ramp gets steeper when I make $H$ larger (without changing $V$)?

*I wonder* if a ramp gets steeper if I make $V$ larger (without changing $H$)?

*I wonder* if I can combine $V$ and $H$ into one number to describe the steepness?

> Experiment! Add, subtract, multiply, and divide $V$ and $H$. Do ramps with the same number have the same steepness? Do steeper ramps have bigger numbers than flatter ramps?

*I notice* that $V \div H$ (or $\dfrac{V}{H}$) tells me something important about a ramp.

Look at this 5 by 3 ramp!

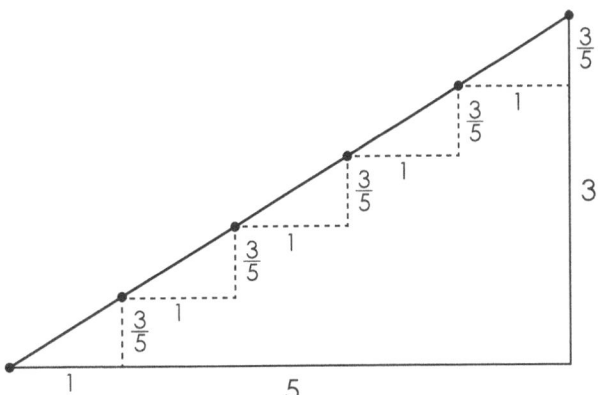

*I wonder* what $\dfrac{H}{V}$ tells me about a ramp?

# SOLUTIONS FOR #2

*From flattest to steepest*: 5 by 3, 13 by 8, 8 by 5

*A strategy with pictures/graphs*:

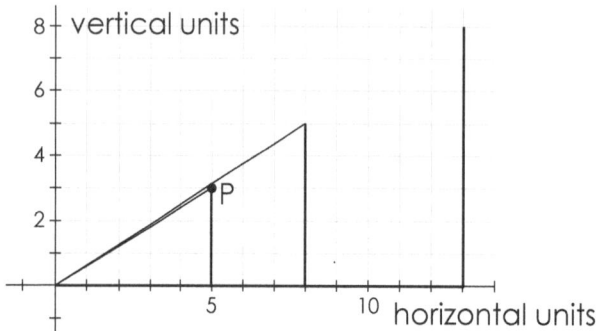

The picture shows that the 8 by 5 ramp is *slightly* steeper than the 5 by 3 ramp. The 13 by 8 ramp (which is not completed, because it makes the picture too hard to read) lies between these! Can you see why? *Hint*: Use your imagination! What happens if you move the lower left tip of the 8 by 5 ramp to point $P$ in the picture?

*A strategy with tables (comparing 5 by 3 to 8 by 5)*:

| Horizontal Units | 0 | 5 | 10 | 15 | 20 | 25 | | 0 | 8 | 16 | 24 |
|---|---|---|---|---|---|---|---|---|---|---|---|
| Vertical Units | 0 | 3 | 6 | 9 | 12 | 15 | | 0 | 5 | 10 | 15 |

                          5 by 3 ramp                       8 by 5 ramp

Look at the last columns in which both vertical lengths are both 15 units. The 8 by 5 ramp has the smaller horizontal length, which makes it steeper.

*A strategy with calculations*:

   » The 5 by 3 ramp climbs $3 \div 5 = 0.6$ or $\dfrac{3}{5}$ units per 1 unit of $H$.

   » The 13 by 8 ramp climbs $8 \div 13 \approx 0.615$ or $\dfrac{8}{13}$ units per 1 unit of $H$.

   » The 8 by 5 ramp climbs $5 \div 8 = 0.625$ or $\dfrac{5}{8}$ units per 1 unit of $H$.

The greater the climb per unit horizontal length, the steeper the ramp!

# CLASSROOM VIGNETTE FOR #2

Ms. Rodriguez's students have had success using pictures, tables, and calculations to order ramps by steepness in Problem #2. Before continuing, she wants them to begin thinking about how to use a single number (the slope) to describe steepness. She draws four ramps on the board:

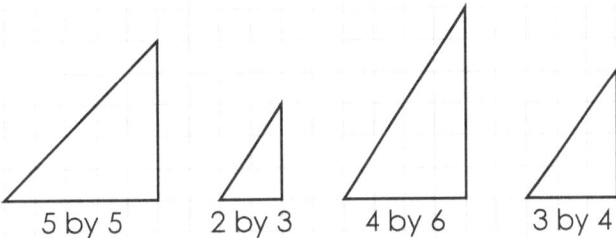

5 by 5    2 by 3    4 by 6    3 by 4

**Ms. Rodriguez:** Can we combine $V$ and $H$ into a single number that describes how steep the 4 by 6 ramp is?

*(Students take turns suggesting numbers: 10, 24, 2, 1.5, and $\frac{2}{3}$ .)*

**Ms. Rodriguez:** Can you explain how you got these numbers?

**Jake:** We used adding, multiplying, subtracting, and dividing.

**Ms. Rodriguez:** Does it matter which operation you use?

**Ellen:** I don't think that adding is good, because the 5 by 5 and the 4 by 6 ramps would have the number 10, but the 4 by 6 is steeper.

**Javier:** I agree with Ellen, because 2 by 3 and 4 by 6 are the same steepness, but you get different numbers when you add them. And bigger ramps would have bigger numbers even if they weren't as steep.

**Ms. Rodriguez:** So if ramps have the same steepness, they should have the same number? *(Students nod.)* Is there another operation that would do this?

**Jake:** Subtracting might work, because $5 - 5 = 0$ and that is the flattest ramp.

*(Students note that other ramps [not on the board] are flatter than 5 by 5, that the order of the numbers matters when you subtract, and that you might get negative answers. Ms. Rodriguez brings them back to the original question.)*

**Ms. Rodriguez:** Javier said earlier that the 2 by 3 and 4 by 6 ramps should have the same number. Does subtraction do that? *(Students shake their heads.)* How do you know?

**Javier:** $3 - 2$ is 1, but $6 - 4$ is 2.

**Ms. Rodriguez:** Can we find an operation that does work?

**Ellen:** When we compared ramps in Problem #2, we multiplied and divided. Maybe one of these works.

**Javier:** Multiplying is no good either. $6 \cdot 4$ is bigger than $3 \cdot 2$, because it's a bigger ramp.

**Jake:** I agree with Javier, but division is not good either, because you can't change the order of the numbers when you divide.

**Ms. Rodriguez:** It sounds like you're saying that our operation should be commutative, which division is not. Is that true, Jake?

**Jake:** Commutative is when you can change the order? *(Ms. Rodriguez nods.)* Yeah.

**Ms. Rodriguez:** Do you agree with Jake that our operation should be commutative?

**Javier:** Yes, because division's like subtraction. It's confusing if you get two answers.

**Ellen:** Wait a minute! What if you tip the ramps so $H$ and $V$ are switched? Maybe that's what the two answers could be for.

**Ms. Rodriguez:** *(Noticing that other students appear to confused.)* Can you say more about that, Ellen?

**Ellen:** If you tip the 4 by 6 ramp into a 6 by 4 ramp so 6 is on the bottom.

*(Ms. Rodriguez draws the new ramp below the 4 by 6 ramp.)*

**Ellen:** See? So that changes the steepness.

**Jake:** I get it! The two numbers are for the two ramps!

**Ms. Rodriguez:** Can someone else put that in their own words?

**Ellen:** $6 \div 4$ is for one of those ramps, and $4 \div 6$ is for the other one.

**Ms. Rodriguez:** What do you get when you divide?

**Javier:** I get 1.5 for $6 \div 4$.

**Ellen:** I get $\dfrac{2}{3}$ for $4 \div 6$, because it's $\dfrac{4}{6}$.

**Jake:** I just noticed something! You get $\dfrac{2}{3}$ for $2 \div 3$ and $4 \div 6$, so division does give you the same answer for both ramps!

**Ms. Rodriguez:** What about $6 \div 4$ and $3 \div 2$?

**Javier:** They're both 1.5!

**Ms. Rodriguez:** How did you figure those out?

**Javier:** Well, 3 is 2 plus half of 2, and 6 is 4 plus half of 4.

**Ms. Rodriguez:** So division gives us the same answer for both ramps no matter which order we divide in. For the 4 by 6 and the 6 by 4 ramps, can we figure out which number goes with which ramp?

**Ellen:** Like we said before, the bigger number goes with the steeper ramp. 4 by 6 is steeper, so I think its number is 1.5.

*(As the discussion continues, the class explores examples with other ramps to test their ideas. Ms. Rodriguez leads them to notice that the steepness number is always $V \div H$*

[not $H \div V$], and connects back to the discussion of rates in Problem #2. Her goal is for students to understand and explain why division gives the same answer for ramps of the same steepness. She makes a note to refer to this discussion in Problem #3 when they decide which quantity to put on the vertical axis of their graphs for green versus blue paint.)

# Problem #3

You are choosing between four mixtures of blue-green paint.

- A. 4 parts green : 8 parts blue
- B. 6 parts green : 9 parts blue
- C. 3 parts green : 5 parts blue
- D. 9 parts green : 14 parts blue

## Directions

- Determine which mixture will look the greenest.
- Explain your answer at least two ways. Use a combination of tables, graphs, diagrams, or calculations.

    Suggestion: Save mixture D for later. (You may decide to use a different thinking strategy for it.)

## Testing the Waters

Solve Problem #3 for mixtures B and C only.

# CONVERSATION STARTERS FOR #3

*What do you notice? What do you wonder?*

*I wonder* if comparing color mixtures is anything like comparing ramps?

*I wonder* if it will help to subtract the green and blue parts?
> Compare 1 part green : 2 parts blue to 99 parts green : 100 parts blue. They have the same difference. Do they have the same "greenness"?

*I notice* that comparing mixtures is easier when I have the same number of parts of green or blue in each mixture.

*I notice* that it helps to know how many parts of green there are for every 1 part blue.

*I wonder* if it matters whether I compare the green to the blue or to the whole mixture?
> Try it both ways!

*I wonder* if it matters which color goes in the numerator when I use fractions?
> Try it both ways! What happens to the size of the fraction when the mixture gets greener?

*I wonder* if it matters which color I put on each axis of my graph?
> Try it both ways! Notice how it affects the steepness of the two graphs.

*I notice* that all of my tables extend back to 0 parts green and 0 parts blue.

*I notice* that something interesting happens when I combine mixtures B and C!

# SOLUTIONS FOR #3

Mixture B is the greenest of the first three mixtures. (We will save D until the end.)

*A strategy using tables*:

| A | |
|---|---|
| **Green** | **Blue** |
| 1 | 2 |
| 2 | 4 |
| 3 | 6 |
| 4 | 8 |

| B | |
|---|---|
| **Green** | **Blue** |
| 2 | 3 |
| 4 | 6 |
| 6 | 9 |
| 8 | 12 |

| C | |
|---|---|
| **Green** | **Blue** |
| 3 | 5 |
| 6 | 10 |
| 9 | 15 |
| 12 | 20 |

Look for entries in which two mixtures have the same amount of one color.

» Mixture B is greener than mixture A, because mixture B has 4 parts green for every 6 parts blue, but mixture A has 3 parts green for every 6 parts blue.

» Mixture B is greener than mixture C, because mixture B has 6 parts green for every 9 parts blue, but mixture C has 6 parts green for every 10 parts blue.

*A strategy using calculations*:

» In mixture A, the amount of green is always $\frac{1}{2}$ of the amount of blue.

» In mixture B, the amount of green is always $\frac{2}{3}$ of the amount of blue.

» In mixture C, the amount of green is always $\frac{3}{5}$ of the amount of blue.

The greatest of these fractions is $\frac{2}{3}$ for mixture B.

*Another strategy using calculations*: Compare the number of green parts to the *total* number of parts.

» Mixture A = 4 parts green : 12 parts total. Green is $33.\overline{3}\%$ of the total.
» Mixture B = 6 parts green : 15 parts total. Green is 40% of the total.
» Mixture C = 3 parts green : 8 parts total. Green is 37.5% of the total.

*A strategy using graphs:*

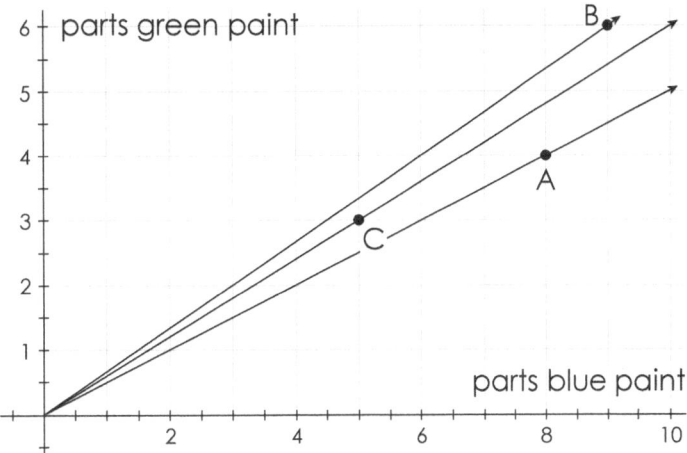

Mixture B has the steepest graph, so it is the most green. (Ask students to compare these graphs using what they learned about steepness of ramps. What happens if the vertical axis represents "parts blue paint"? Try it!)

The "greenness" of mixture D is between B and C. If you pour mixtures B and C together, you will get mixture D! It will be greener than C and less green than B. You may use fractions to stand for the strengths of green compared to blue:

$$\frac{3}{5} < \frac{9}{14} < \frac{6}{9}$$

# STAGE 2

In Problem #4, students extend their understanding of ratios and rates to make paint mixtures satisfying specific conditions. As always, be sure to encourage them to create, share, and compare strategies that make sense to them.

Problem #5 is a bit different. You do not need to do it in order to understand the later problems, but it is interesting to try if you have time! It explores a common error that students make when they add fractions. This leads to some interesting discoveries about comparing fractions. Students' experience with the ramps and paints might give them common sense ways to understand why the fractions are increasing in size.

## *What Students Should Know*

- » Ratio and rate concepts from Stage 1.

## *What Students Will Learn*

- » Use *tape diagrams* to represent and solve problems.
- » Use part-whole comparisons in problem-solving strategies.
- » Use knowledge gained from ratios and rates to understand how the value of $\frac{a+c}{b+d}$ compares to $\frac{a}{b}$ and $\frac{c}{d}$.

# Problem #4

- A: 4 parts green : 8 parts blue
- B: 6 parts green : 9 parts blue
- C: 3 parts green : 5 parts blue

You want to make 20 pints of each paint mixture.

## *Directions*

- Calculate the amount of blue and green paint needed for each mixture. Explain your thinking.
- Do the mixtures in any order you like.

### *Testing the Waters*

Solve Problem #4 for mixture B only.

# CONVERSATION STARTERS FOR #4

*What do you notice? What do you wonder?*

This is a *tape diagram* for mixture C.

| green | | | |
|---|---|---|---|

| blue | | | | |
|---|---|---|---|---|

*I notice* that the tape diagram helps me visualize the parts and wholes.

*I notice* that the tape diagram helps me think about percentages.

*I wonder* how I can use the tape diagram to solve the problem?
Suggestion: Fill in each rectangle with some number of pints per part.

# SOLUTIONS FOR #4

Mixture A $= 6\frac{2}{3}$ pints green : $13\frac{1}{3}$ pints blue

Mixture B = 8 pints green : 12 pints blue

Mixture C = 7.5 pints green : 12.5 pints blue

*A solution for mixture C Using a tape diagram:*

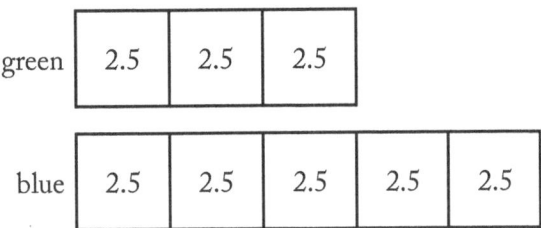

For every 3 parts green, there are 5 parts blue. Because there are 8 parts in the entire 20 pints, each part contains $20 \div 8 = 2.5$ pints. Therefore, there are $2.5 \times 3 = 7.5$ pints of green paint and $2.5 \times 5 = 12.5$ pints of blue paint.

Check: $7.5 + 12.5 = 20$ pints

*Another solution for Mixture C:*

Green is $\frac{3}{8}$ (37.5%) of the total.      $\frac{3}{8}$ (37.5%) of 20 pints = 7.5 pints

Blue is $\frac{5}{8}$ (62.5%) of the total.      $\frac{5}{8}$ (62.5%) of 20 pints = 12.5 pints

# Problem #5

This list was created using some ideas from Problems #1–4.

$$\frac{3}{5}, \frac{11}{18}, \frac{8}{13}, \frac{13}{21}, \frac{5}{8}$$

## Directions

- Show how the list was created.
- Make the list longer by continuing the process.
- Compare the sizes of the fractions in your list.
- Explain what happens and why.

## Diving Deeper

- Make other lists like the one in Problem #5. Compare and contrast your lists.

- How does each number sentence follow from the one before it? How could you use this idea in reverse to compare fractions like $\frac{5}{7}$ and $\frac{13}{18}$?

- $\frac{1}{2} < \frac{3}{5}$  $\qquad$ $2\frac{1}{2} < 2\frac{3}{5}$  $\qquad$ $\frac{2}{5} > \frac{5}{13}$  $\qquad$ $1\frac{2}{5} > 1\frac{5}{13}$  $\qquad$ $\frac{5}{7} < \frac{13}{18}$

# CONVERSATION STARTERS FOR #5

*What do you notice? What do you wonder?*

*I notice* that the list has some of the same fractions as the problems about ramps and paints.

*I notice* that I can make a lot of addition or subtraction equations with the numerators and denominators of these fractions.

*I notice* that the smallest numerators and denominators are on the ends of the list, and the largest ones are in the second and fourth spots on the list.

*I notice* that something interesting happens when I remove the second and fourth fractions from the list.

*I notice* that the fractions are very close in size!

*I notice* that the fractions increase from left to right.

*I wonder* why the fractions do this? Will it keep happening if I extend the list?

*I wonder* what will happen if I draw a ramp for each fraction?

*I notice* that finding the fraction between two consecutive fractions in the list is a lot like combining paint mixtures B and C to create mixture D in Problem #3.

# SOLUTIONS FOR #5

*Step 1*: Start with the first and last fractions.

$$\frac{3}{5} \quad \frac{5}{8}$$

*Step 2*: Find the fraction between them by adding the numerators and adding the denominators.

$$\frac{3}{5} \quad \boxed{\frac{8}{13}} \quad \frac{5}{8}$$

*Step 3*: Use the same process to insert fractions between the first and last pairs.

$$\frac{3}{5} \quad \boxed{\frac{11}{18}} \quad \frac{8}{13} \quad \boxed{\frac{13}{21}} \quad \frac{5}{8}$$

» Insert the correct fraction between each of these to make a list of nine fractions!

$$\frac{3}{5} \quad \boxed{\frac{14}{23}} \quad \frac{11}{18} \quad \boxed{\frac{19}{31}} \quad \frac{8}{13} \quad \boxed{\frac{21}{34}} \quad \frac{13}{21} \quad \boxed{\frac{18}{29}} \quad \frac{5}{8}$$

» The fractions increase from left to right. (This is easier to see if you write them as decimals. They are all very close!)
» This increase occurs when you add the numerators and the denominators of two unequal fractions, because the value of the resulting fraction is between them! (To understand why, see what happens when you combine paint mixtures B and C in Problem #3.)

*I wonder* what will happen if I start with two equal fractions?

*I wonder* if I can extend the list to the right of $\frac{5}{8}$?

# STAGE 3

In Stage 3, students analyze data from a science experiment to learn why hot air balloons rise. It is fun to ask them for their own ideas about this before beginning!

The two problems in Stage 3 are fairly sophisticated, so I like to give students more experience solving problems with familiar rates first (miles per gallon, dollars per ounce, etc.). I just adapt problems from a math textbook, ensuring that students use a blend of diagrams, tables, graphs, and calculations to discuss and solve them. Depending on how students are progressing, I sometimes wait until later in the year to do the problem.

To get started with Problem #6, define a *proportional relationship* as a set of equivalent ratios (ratios that have the same rate). For example, the relationship between $H$ and $V$ for ramps with the same steepness is proportional, because $V \div H$ always has the same value. Ask your students to look at tables and graphs from their text or from Stages 1 and 2 to find examples and nonexamples of proportional relationships.

## *What You Will Need*

  » Graph paper and calculators.

## *What Students Should Know*

  » Use diagrams, tables, graphs, and calculations to solve ratio problems.
  » mL is the metric abbreviation for *milliliters* (1 mL = 0.001 liters).

## *What Students Will Learn*

  » Understand that in a *proportional relationship*, each ratio has the same rate.
  » Write formulas for proportional relationships.
  » Understand that real-world proportional relationships may be approximate.
  » Know that graphs of proportional relationships are lines that contain (0,0).
  » Understand that if you add or subtract a number from a quantity in a proportional relationship, it will not remain proportional.

# Problem #6

The table shows data collected for a small sample of air that was gradually heated. The amount of air and its pressure were kept constant.

| | Temperature and Volume of Air (Constant Pressure) | | | | | |
|---|---|---|---|---|---|---|
| Temperature (°K) | 266 | 273 | 301 | 320 | 345 | 355 |
| Volume (mL) | 4.5 | 4.6 | 5.1 | 5.4 | 5.8 | 6.0 |

## Directions

- Determine if the relationship between temperature and volume is proportional*. Explain your reasoning.
- Graph the data, and use it to support your conclusion.
- Create an equation showing the relationship between the temperature and volume. Explain the meaning of all variables and number(s) in your equation.
- Explain why hot air balloons rise. Use what you learned above.

    *In a *proportional relationship*, every output : input ratio has the same rate.

# CONVERSATION STARTERS FOR #6

*What do you notice? What do you wonder?*

*I wonder* what °K means?

°K means "degrees Kelvin." Scientists often use the Kelvin temperature scale. It is related to the Celsius scale by the formula $C = K - 273$.

*I wonder* why the volume increases when the temperature rises?

The volume of the air increases, because the molecules move faster when the temperature rises. The amount of air does not change!

*I wonder* what these temperatures are in Celsius or Fahrenheit units?

Find out! Use the formulas $C = K - 273$ and $F = 1.8 \cdot C + 32$.

*I wonder* how exact the measurements are?

They are as exact as the smallest place value. (No measurement is ever perfect.)

*I notice* that the temperature (and volume) values in the table increase by a different amount each time.

*I wonder* what will happen if I compare volumes (or temperatures) to each other?

Try it! For example, compare $301 \div 273$ and $5.1 \div 4.6$.

*I notice* that the volume : temperature ratios are not equivalent, but they are close!

*I wonder* if the rates must be exactly equal for the relationship to be proportional?

Scientists consider the relationship between volume and temperature in this situation to be proportional, even though the calculated rates are not exactly equal. Remember that measurements are never exact.

*I wonder* which quantity should go on the vertical axis of my graph?

*I wonder* if the points on my graph will form a line?

# SOLUTIONS FOR #6

The relationship is approximately proportional, because all pairs of values are nearly *equivalent ratios*, meaning that they have about the same *rate*.

$4.5 \div 266 \approx 0.01692$ mL per 1°K        $4.6 \div 273 \approx 0.01685$ mL per 1°K

$5.1 \div 301 \approx 0.01694$ mL per 1°K        $5.4 \div 320 \approx 0.01688$ mL per 1°K

$5.8 \div 345 \approx 0.01681$ mL per 1°K        $6.0 \div 355 \approx 0.01690$ mL per 1°K

The rate is the increase in volume when the temperature increases by 1°K. (If students divide in the opposite order, the rate shows the increase in number of degrees needed to raise the volume by 1 mL (just over 59°K per mL). The two rates are reciprocals!)

The data lie nearly on a line, meaning that the volume increases at about a constant rate, which must happen in a proportional relationship. In Problem #7, students will learn that the graph has one other important feature!

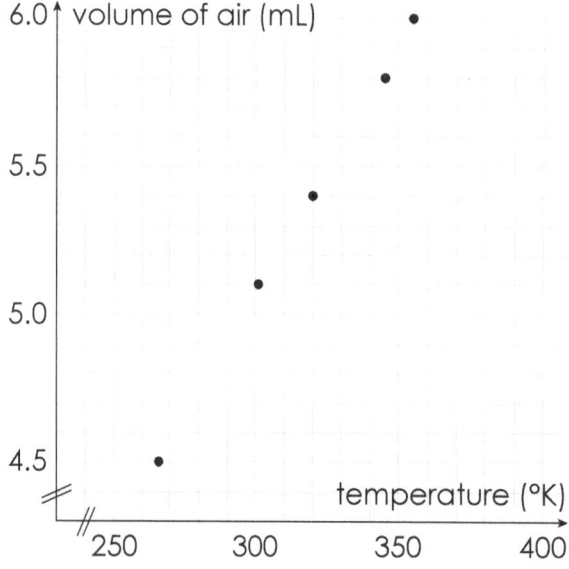

*Possible formula*: $V \approx 0.0169 \cdot T$ . $V$ is the volume (mL) and $T$ is the temperature (°K). 0.0169 is the *unit rate* for the proportional relationship!

A hot air balloon rises because the molecules move faster when the air is heated, causing the volume of the air in the balloon to increase. Because the same amount of air is occupying a larger space, it is less dense. Eventually, the entire balloon weighs less than the air it displaces, causing it to rise!

*I wonder* why scientists often use the Kelvin scale instead of the Celsius scale?

# Problem #7

The pattern in this table may be extended to lower temperatures.

| | Temperature and Volume of Air (Constant Pressure) | | | | | |
|---|---|---|---|---|---|---|
| Temperature (°K) | 266 | 273 | 301 | 320 | 345 | 355 |
| Volume (mL) | 4.5 | 4.6 | 5.1 | 5.4 | 5.8 | 6.0 |

## Directions

- Estimate the volume when the temperature is 0°K. Explain your thinking.
- Draw some conclusions about the meaning of 0°K. Explain.
- Change the temperature units from Kelvin to Celsius, and decide if the relationship is still proportional. Explain.

# CONVERSATION STARTERS FOR #7

*What do you notice? What do you wonder?*

*I wonder* what is the best way to find the volume at 0°K?

Try the formula, table, or graph. The formula is quick. The table takes more work, but you might learn more from it. The graph would have to be extended a long way!

*I wonder* if the formula is still true when the temperature is very low?

## After Students Have Found the Volume

*I wonder* if the volume can actually get down to 0 mL?

*I wonder* what happens if the temperature goes below 0°K?

## After Students Discover that Changing to Celsius Ruins the Proportionality

*I wonder* why the relationship isn't proportional when you use Celsius units?

Think about the ramps in Problems #1 and #2. What happens if you trace your table back so that $H$ is 0? What happens if you subtract 1 (or another number) from all values of $H$ in a ramp's table?

*I notice* that the graphs of all proportional relationships we have seen are straight lines that include the point (0,0).

# SOLUTIONS FOR #7

The volume is 0 mL when the temperature is 0°K!

*A strategy using the formula*: When $T = 0°K$, $V = 0.0169 \cdot 0 = 0$ mL.

*A strategy using a table and calculations*: Begin with $T = 266°K$ and $V = 4.5$ mL. To reach 0°K, $T$ must decrease by 266°K. Because each decrease of 1°K results in a 0.0169 mL decrease in the volume, the total decrease in volume is approximately $266 \cdot 0.0169 \approx 4.5$ mL. Because the volume began at 4.5 mL, it will decrease to 0 mL! If you continue the graph to the left, it will eventually touch the *origin* [the point (0,0)].

Because the volume cannot be negative, students may conclude that if the temperature drops below 0°K, the volume will remain 0 mL. In fact, the temperature *cannot* drop below 0°K! 0°K is *absolute zero*, the coldest possible temperature. Some students may be interested in doing more research on this topic!

To change the temperatures to Celsius units, just subtract 273 from each.

| Temperature (°C) | -7 | 0 | 28 | 47 | 72 | 82 |
|---|---|---|---|---|---|---|
| Volume (mL) | 4.5 | 4.6 | 5.1 | 5.4 | 5.8 | 6.0 |

The ratios are no longer equivalent! For example, $4.5 \div -7$ has a negative value, $5.1 \div 28$ and $5.4 \div 47$ have different values, and $4.6 \div 0$ is not even a number! Therefore, it is not a proportional relationship, even though the physical meaning of the data has not changed! This illustrates an important feature of proportional relationships: their tables and graphs must begin at (0,0). If students look back at earlier problems, they will see that all of the tables can be traced back to (0,0).

## ALGEBRA CONNECTIONS

Even though you may not see a lot of variables in this exploration, it has huge connections to algebra. As students explore the relationship between rates of change and the steepness of ramps, they are setting the stage for a deeper understanding of slope in algebra courses. The Classroom Vignette suggests some ideas for digging more deeply into these connections, but even if you do not have these discussions, the experience your students gain from looking closely at the relationships between horizontal and vertical sides and the steepness of the ramps is invaluable.

More broadly, the whole idea of proportional relationships that students begin to explore in this activity is foundational for understanding algebra. It will appear in the study of proportions, linear functions, and properties and procedures for working with fractions. It is conceptually at the heart of many of the problems they will solve throughout their math careers.

Problem #5 offers some specific applications of algebra for students who have the background and are interested. Encourage them to explore the relationships between $\frac{a}{b}$, $\frac{c}{d}$, and $\frac{a+b}{c+d}$. When the first two fractions are equal, all three are equal! When the first two fractions are not equal, the third fraction is always between the first two. All students may explore this by substituting a lot of numbers and looking for patterns. Algebra students may challenge themselves to prove these results.

# Exploration 3

## Gear Up!

Bicycle gears offer a great real-world connection to ratios and rates! Your students may benefit from looking at a bike (or a picture of the gears) as you introduce this activity. Show them that the gears consist of two sets of toothed disks called *sprockets*. One set is attached to the pedal and the other to the rear wheel. The pedal often has two or three sprockets. The wheel usually has more.

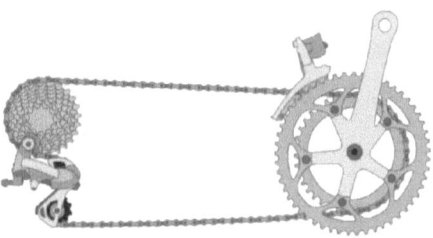

Drawing by Keith Onearth, http://babs.co/gears/wikipedia_53_39_11_25_large.png

When you change gears, the chain moves between different sprockets. If you are looking at a real bike, have students count the teeth on the sprockets that the chain is attached to. Ask them to predict how many times the rear wheel will turn each time the pedal turns once. Then test it to see what happens!

For example, if the pedal sprocket has 39 teeth and the wheel sprocket has 13, the wheel will turn 3 times for every turn of the pedal. Explore a few different gear combinations. Students should notice two things:

» The smaller sprocket turns faster than the larger one.
» You divide the pedal sprocket teeth by the wheel sprocket teeth to predict the number of wheel turns per pedal turn. (This is called the *gear ratio*.)

The CCSS use the word *ratio* for comparisons expressed as two numbers and the word *rate* for single-number (per-unit) comparisons. This distinction is helpful when students first learn the concepts, and I try to follow it in this book. However, in more advanced math courses and in real-world situations, *ratio* is often used for single-number comparisons, especially when the quantities have the same type of unit. For example, a *gear ratio* such as 39 : 13 is usually expressed as a single number (3, in this case).

71

In Stage 1, students explore a bicycle gear assembly having two pedal sprockets and 10 wheel sprockets. They calculate the gear ratios and use them connect the math to the real-world situation.

To get started, share the information in the introduction to this exploration. Tell students that the higher gears on a bike have a larger gear *ratio*. Many of them may know from experience that bikes in higher gears will go faster, but will be harder to pedal.

Problem #2 contains many important algebraic ideas. You can ensure that students get the most out of it by helping them recognize proportional relationships as they appear in tables, graphs, and formulas.

## What You Will Need

> » A bicycle (or a picture of bicycle gears).

## What Students Should Know

> » Calculate rates from ratios.
> » Create tables, graphs, and formulas from ratios.

## What Students Will Learn

> » Use ratios and rates to describe and analyze real-world situations.
> » Understand that in a *proportional relationship*, each ratio has the same (per-unit) rate.
> » Recognize a *proportional relationship* and its rate from a table, graph, and formula.
> » Understand that the graph of a *proportional relationship* is a line beginning at $(0, 0)$.

# Problem #1

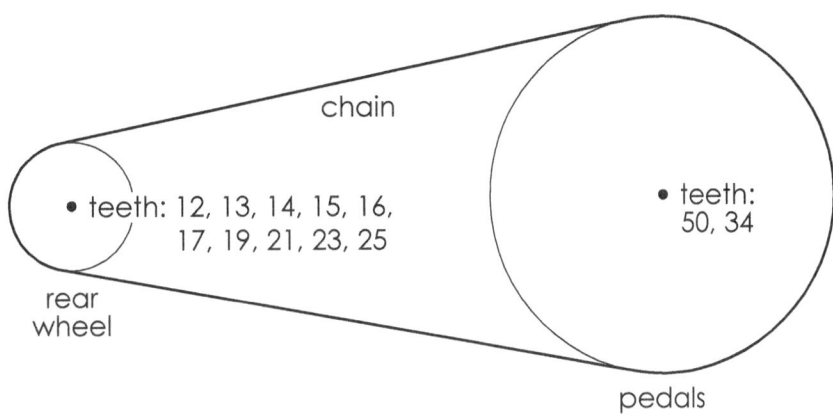

chain

• teeth: 12, 13, 14, 15, 16, 17, 19, 21, 23, 25

• teeth: 50, 34

rear wheel

pedals

**Gears on Jamilah's Bike**

The manufacturer advertises this as a 20-speed bike.

## Directions

- Order the gears from low to high. Show your calculations, and explain your thinking.
- Evaluate the claim that it is a 20-speed bike. Explain your thinking.
- Make some observations about how the gear ratios affect the way the bike rides.

### *Diving Deeper*

Can you tell what gear the bike is in without seeing the teeth? (Measure the circles!)

# CONVERSATION STARTERS FOR #1

*What do you notice? What do you wonder?*

*I notice* that there are 20 gear combinations, 10 each for the two pedal sprockets.

*I notice* that the wheel always turns faster than the pedal, because its sprocket is smaller.

*I notice* that some of the gear ratios are almost (or exactly) the same.

*I wonder* how the manufacturer chose this particular set of gears?

*I wonder* why the teeth on the wheel sprocket increase by two when there are more teeth?

*I wonder* what the gears are on my bike?

# SOLUTIONS FOR #1

To calculate the gear ratio, divide the number of teeth on the pedal sprocket by the number of teeth on the wheel sprocket. For example, the gear ratio for 50:12 is $50 \div 12 \approx 4.17$. The list below shows that some gears in the middle have equal or nearly equal gear ratios. It is probably more accurate to describe the bike as having 16 speeds.

| Gear | Gear Ratio | Gear | Gear Ratio |
|------|-----------|------|-----------|
| 34:25 | 1.36 | | |
| 34:23 | 1.48 | | |
| 34:21 | 1.62 | | |
| 34:19 | 1.79 | | |
| 34:17 | 2.00 | 50:25 | 2.00 |
| 34:16 | 2.13 | 50:23 | 2.17 |
| 34:15 | 2.27 | | |
| 34:14 | 2.43 | 50:21 | 2.38 |
| 34:13 | 2.62 | 50:19 | 2.63 |
| 34:12 | 2.83 | | |
| | | 50:17 | 2.94 |
| | | 50:16 | 3.13 |
| | | 50:15 | 3.33 |
| | | 50:14 | 3.57 |
| | | 50:13 | 3.85 |
| | | 50:12 | 4.17 |

*Observations*: The gear ratio equals the number of times the wheel turns when the pedal turns once. High gear ratios make the bike go faster, but they make it harder to turn the pedals. You can increase a gear ratio by increasing the number of teeth on the pedal sprocket or decreasing the number of teeth on the wheel sprocket.

# Problem #2

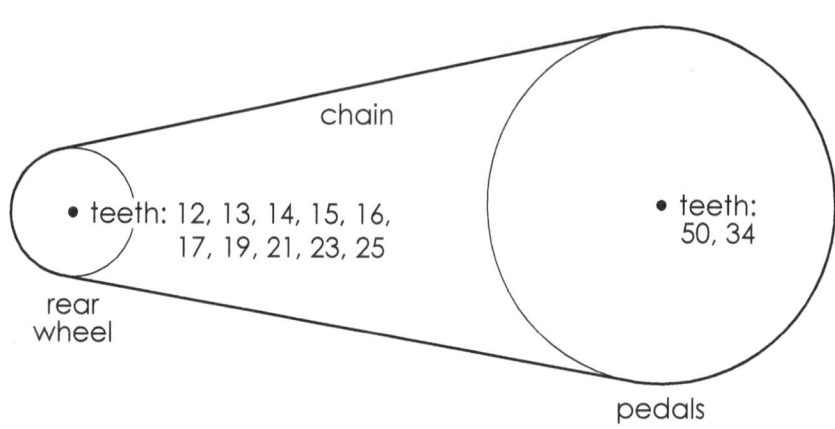

rear
wheel

chain

• teeth: 12, 13, 14, 15, 16,
17, 19, 21, 23, 25

• teeth:
50, 34

pedals

**Gears on Jamilah's Bike**

Tables, graphs, and formulas can help you understand this situation better.

## *Directions*

For each gear, 50:12 and 34:25:

- Make a table comparing wheel turns to pedal turns.
- Draw a graph comparing wheel turns to pedal turns.
- Write an equation for calculating wheel turns from pedal turns.
- Calculate a rate. Relate it to the table, graph, and formula. Explain your thinking.
- Decide if the relationship is proportional. Justify your reasoning.

## *Testing the Waters*

Follow the directions for a bike with gears 44:11 and 44:22.

# CONVERSATION STARTERS FOR #2

*What do you notice? What do you wonder?*

*I wonder* which variable should be the input?
> Hint: you are probably predicting the number of wheel turns from the number of pedal turns.

*I notice* that the gear ratio is helpful when I create the table.

*I notice* that one row of my table looks like the gear numbers in reverse.

*I notice* that the wheel turns 50 times for every 12 pedal turns with a 50:12 gear ratio.

*I wonder* if the wheel turns $a$ times for every $b$ pedal turns with an $a$:$b$ gear ratio?
> Test the idea for other gear ratios! (The answer is yes. Explain why!)

*I wonder* if I should draw the two graphs in the same or different coordinate grids?
> They are easier to compare if you draw them in the same coordinate grid.

*I notice* that the input of 1 is important in my table and graph.
> The output for an input of 1 is the rate (and the gear ratio) of the relationship.

*I notice* that both graphs are straight lines.

*I notice* that the graph for the larger gear ratio is steeper.
> This is true when you choose the pedal turns for the input.

*I notice* that both tables and graphs start at $(0, 0)$.

*I wonder* what would happen if I chose wheel turns for the input?

# SOLUTIONS FOR #2

*Tables for pedal turns (input) and wheel turns (output):*

| 50:12 | |
|---|---|
| Pedal Turns | Wheel Turns |
| 0 | 0.00 |
| 1 | 4.17 |
| 2 | 8.33 |
| 3 | 12.50 |
| 4 | 16.67 |
| 5 | 20.83 |
| 6 | 25.00 |
| 7 | 29.17 |
| 8 | 33.33 |
| 9 | 37.50 |
| 10 | 41.67 |
| 11 | 45.83 |
| 12 | 50.00 |

| 34:25 | |
|---|---|
| Pedal Turns | Wheel Turns |
| 0 | 0.00 |
| 1 | 1.36 |
| 2 | 2.72 |
| 3 | 4.08 |
| 4 | 5.44 |
| 5 | 6.80 |
| 6 | 8.16 |
| 7 | 9.52 |
| 8 | 10.88 |
| 9 | 12.24 |
| 10 | 13.60 |
| 11 | 14.96 |
| 12 | 16.32 |

*Graphs for pedal turns (input) and wheel turns (output):*

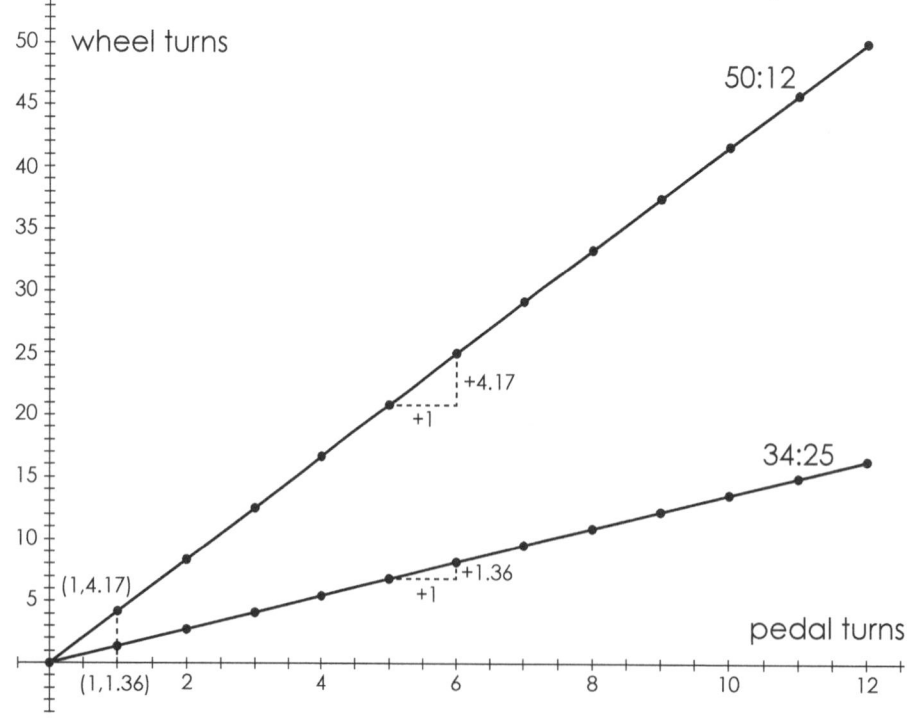

*Formulas for the pedal turns (P) and wheel turns (W):*

$50{:}12$ gear ratio: $W \approx 4.17 \cdot P$    $34{:}12$ gear ratio: $W = 1.36 \cdot P$

*Rates for pedal turns compared to wheel turns:*

The rate for the 50:12 gear is $50 \div 12 \approx 4.17$ wheel turns per pedal turn.

The rate for the 34:25 gear is $34 \div 25 = 1.36$ wheel turns per pedal turn.

The rates are the gear ratios! In the table and the formula, you multiply the input (pedal turns) by the rate to get the output (wheel turns). The rate is also the increase in the number of wheel turns each time the number of pedal turns increases by 1.

Both graphs are straight lines, because the number of wheel turns changes at a constant rate. The greater the rate, the steeper the line. The rates are the output value when the input equals 1 (see the table and the graph). The small triangles on the graph show that it rises by the rate every time you move 1 unit to the right.

Both relationships are proportional, because the rate for every input/output pair is the same within each table. For example, in the 34:25 table, $4.08 \div 3$ and $9.52 \div 7$ both equal 1.36. You can also see that the relationships are proportional by noticing that the graphs are straight lines that include the *origin* (0,0).

# STAGE 2

Now that students have analyzed the gears on Jamilah's bike, they are ready to solve a challenging problem! They may need some help at first making a connection between the size of the wheel and the forward motion of the bike. They can experiment by rolling lids or other circular objects on a flat surface. Let them decide what measurements to make and what calculations to do. They should discover that for each full turn of the lid (wheel), it moves forward a distance equal to its circumference.

A vocabulary note: A cyclist's *cadence* is rate at which she or he is pedaling. It is often expressed in revolutions per minute (*rpm*).

## What You Will Need

>   »   Lids or other circular objects (recommended).

## What Students Should Know

>   »   Understand gear ratios from Stage 1.
>   »   Convert between units of measurement (both customary and metric).
>   »   Find the circumference of a circle (circumference = $\pi$ · diameter).

## What Students Will Learn

>   »   Use math to analyze and understand circular and straight-line motion.
>   »   Solve challenging problems using ratios and rates.
>   »   Interpret results of calculations.

# Problem #3

Jamilah has a new job 10 miles from her home. She hopes to be able to bike there in about 25 minutes. She has 700 mm wheels on her bike (including the tires), and she knows that 80 rpm is an ideal pedaling cadence for her.

## Directions

- Find an approximate gear ratio that Jamilah needs. Explain your thinking.
- Decide if you think this gear ratio is practical for her. Explain your thinking.
- If you think it is not practical, make some recommendations for her.

# CONVERSATION STARTERS FOR #3

*What do you notice? What do you wonder?*

*I wonder* what Jamilah's speed will be in miles per hour?

*I wonder* how far the bike moves forward for each full turn of the wheel?

    If you are not sure, experiment by rolling something circular on the floor. Make some measurements, and do some calculations.

*I wonder* what happens if the wheel slips or skids?

*I wonder* how I can find a gear ratio without knowing the gear?

    What does the gear ratio tell you about the rates at which the pedal and the wheel turn?

*I wonder* how many millimeters are in one mile?

    You may have to look up some conversion factors.

## After the Gear Ratio is Calculated

*I wonder* if Jamilah could increase her speed by getting a bike with a larger wheel?

    It may be expensive, and the larger wheel would make it harder to turn the pedal, just like a higher gear does!

# SOLUTIONS FOR #3

Jamilah needs a gear ratio of about 3.66. Notice that 10 miles in 25 minutes is $10 \div 25 = 0.4$ miles per minute, which is 24 mph. This seems like a realistic speed for her to ride.

*One strategy to calculate the gear ratio:*
- » Calculate the number of millimeters that Jamilah wants to travel in one minute.
- » Find the number of times the wheel must turn each minute to make this happen.
- » Divide this number by 80 pedal turns per minute to get the gear ratio.

*Millimeters per minute that Jamilah wants to travel:*
10 miles per 25 min = 0.4 miles per min
0.4 miles per min $= 0.4 \cdot 5280 \cdot 12 = 25,344$ in per min
25,344 in per min $= 25,344 \cdot 2.54 = 64,373.76$ cm per min
64,373.76 cm per min $= 64,373.76 \cdot 10 = 643,737.6$ mm per min

*Number of times the wheel must turn:*
The circumference of the wheel is $700 \cdot \pi$ mm, so the wheel must turn $643,737.6 \div (700 \cdot \pi) \approx 292.726$ times per minute at Jamilah's desired speed.

*Gear ratio:*
292.726 wheel turns per min $\div$ 80 pedal turns per min $\approx 3.66$

The closest gear on her bike is 50:14. This is one of the highest gears, so the pedals may be hard to turn. If the terrain is flat, and if Jamilah is in good condition, it may work for her. If she uses a lower gear, she will have to pedal faster than her ideal cadence, or she will need to leave a little extra time to get to work.

# STAGE 3

In Stage 3, students take the situation with Jamilah's bike to the next level of challenge. Instead of doing calculations to solve a problem, they create a formula that will solve the problem for any choice of values. This time, they are calculating the speed of the bicycle instead of the gear ratio, and they have to decide what the variables should be! Learning to create and test your own formulas is an important skill for using spreadsheets or other technological tools to solve real-world problems.

## What Students Should Know

- » Write algebraic expressions for simple real-world relationships.
- » Understand bicycle gears and gear ratios (see Stages 1 and 2).
- » Understand how the size of a wheel affects the motion of a bicycle.

## What Students Will Learn

- » Identify appropriate variables to use in an equation.
- » Write an algebraic equation for a complex real-world relationship.
- » Test an algebraic equation for reasonableness.

# Problem #4

Jamilah is creating a biking website. She wants to include an online calculator that will find a biker's speed based on information that the user enters.

## Directions

- Choose variables for the equation.
- Create a formula that Jamilah can put into her computer code to do the calculation.
- Test your formula.

# CONVERSATION STARTERS FOR #4

*What do you notice? What do you wonder?*

*I notice* that the variables determine the information that the user enters.

*I notice* that Problem #3 has ideas for variables I can use.

*I notice* that it helps to do a calculation with numbers first.
> You can turn a calculation process into a formula!

*I notice* that there is more than one way to write the formula.

*I notice* that many parts of the formula have to do with converting units of measurement.

*I notice* that some of the steps in the formula can be done in a different order.

*I notice* that some of the steps in the formula can be combined.

*I wonder* if the answers should be rounded?
> Users probably do not care to see too many decimal places.

*I wonder* how simply I can write the formula?

*I wonder* how I can test the formula?
> Check that it gives reasonable answers.
> Check it against your results in Problem #3.
> Compare it to the results given by other students' formulas.
> Look for an online calculator that performs the same task.

*I wonder* what happens to the formula if I use different units?
> For example, what if the diameter were given in inches instead of millimeters?

*I wonder* how the formula will look if I leave out the multiplication symbols and use fractions to show division?
> See the Algebra Connections for ideas.

# SOLUTIONS FOR #4

*A possible set of variables*:

   $P$: teeth on the pedal sprocket
   $W$: teeth on the wheel sprocket
   $C$: cadence (revolutions per minute)
   $D$: diameter of the wheel (millimeters)

*A formula*:

   In miles per hour: $speed = C \cdot (P \div W) \cdot D \cdot \pi \div 10 \div 2.54 \div 12 \div 5280 \cdot 60$
   If you do the calculations at the end, the expression simplifies to:

$$speed \approx C \cdot (P \div W) \cdot D \cdot 0.000117$$

This makes the calculation easier, but it hides the thinking that led to the formula! (See the Algebra Connections to write the formula more compactly.)

*Thinking process*:

   »  Multiply $C$ by the gear ratio $P \div W$ to get number of wheel turns per minute.
   »  Multiply by the circumference $D \cdot \pi$ to get the number of millimeters per minute.
   »  Divide by 10 to convert to centimeters per minute.
   »  Divide by 2.54 to convert to inches per minute.
   »  Divide by 12 and then by 5280 to get the number of miles per minute.
   »  Multiply by 60 to get the number of miles per hour.

*One test for the formula*:

   If you substitute $C = 80$, $P \div W = 3.66$, and $D = 700$ into the formula, the answer is about 24 mph. This is consistent with the results from Problem #3.

# ALGEBRA CONNECTIONS

As students explore three ways to represent mathematical relationships (tables, graphs, and formulas) in Problem #2, they are setting the stage for using variables to solve proportions in more efficient ways. For now, you can support this learning by asking questions such as "If the wheel turns 20.83 times when the pedal turns 5 times, how many times does the wheel turn if the pedal turns 8 times?" Students should be able to justify their answer using any of the three representations.

Problem #2 also prepares students for the study of linear functions in algebra. Their observations about constant rates of change and straight-line graphs in this problem will lead to the study of a formula for the *slope* of a line. When they attend to the values of the output for inputs of 0 and 1, they are preparing to study both the *slope* and the *y-intercept* of a graph. I resist the temptation to give prealgebra students algebraic formulas and procedures for these concepts, because I find that when I teach them prematurely, students use memorization in place of thinking.

The formula in Problem #4 may be written more clearly and compactly by leaving out the multiplication symbols and by using fractions to represent division.

$$speed \approx 0.000117 \frac{CPD}{W}$$

Encourage students to try writing the formula in a form such as this! In the process, they will be forced to grapple with questions about properties of operations. For example, they may discover that $C \cdot \dfrac{P}{W} = \dfrac{C \cdot P}{W}$. Relationships like these are often easier to notice and understand when you write division in terms of fractions.

# Exploration **4**
## Perplexing Percentages

In my experience, talented math students tend to make sense of percentages without much trouble, though they may struggle with challenging problems when they know only one way to think about them. Over time, I have found that students learn more when they solve percentage problems with their own ideas rather than steps I have taught them. Their strategies often reveal interesting insights!

In this exploration, I share some problems that have helped my students deepen their insights and clarify any misconceptions. I usually blend these problems with exercises from a textbook so that students firm up new skills and understandings as they work.

Before beginning the exploration, your students should understand percentages as "parts per 100." They should be able to express halves, thirds, fourths, fifths, and tenths as percentages.

Students who have not have learned quick procedures for percentage calculations (dividing a part by a whole to find a percentage, for example) will develop strategies based on their personal understanding. Those who do know some procedures should still find plenty here to challenge them.

# STAGE 1

Stage 1 contains three problems. Problem #1 deals with basic percentage calculations: finding an unknown percentage, an unknown part, or an unknown whole. If students have not learned quick ways to do these things, guide them to find their own strategies based on what they already know.

> *Note*: Flag measurements for Problem #1 are shown in feet and feet$^2$. In practice, cloth is typically measured by the yard with a width that varies depending on the type of fabric. The fabric used for a U.S. flag depends upon the intended use of the flag. Some students may be interested in exploring further by determining the yard measures required for different types of fabric.

Problems #2 and #3 may be less challenging than #1, even though they involve more advanced content (percentage increase and decrease). They provide a bridge to the more complex questions in Stage 2, and they introduce students to double number lines—a powerful tool for visualizing percentages and other proportionality concepts. If the numbers are causing difficulty, try Testing the Waters first.

Encourage your students to compare calculation strategies. For example, in Problem #3, compare Strategies 1 and 3 (see the Solutions) to notice that $6600 \div 71 = (6600 \div 7100) \cdot 100$ and think about why this makes sense. Discussions like these are great for supporting students' emerging algebraic reasoning!

## What You Will Need

> » Graph paper (helpful for double number lines or other diagrams).

## What Students Should Know

> » Understand the meaning of *area*, and find the area of a rectangle.
> » Understand the meaning of *percent* as "per 100."
> » Be familiar with basic fraction, decimal, and percentage conversions.
> » Understand the meaning of *ratios* and *equivalent ratios*.

## What Students Will Learn

> » Solve problems involving unknown percentages, parts, or wholes.
> » Use double number line diagrams to represent and understand percentages.
> » Solve percentage increase and decrease problems using multiple strategies.

NAME: _____ DATE: _____

# Problem #1

2.2 square feet

## Directions

Do the following tasks in any order:

- Find the area of the red stripes as a percentage of the area of the flag.
- Find the area of the white stripes as a percentage of the area of the flag.
- Calculate the amount of cloth needed to make the red stripes.
- Calculate the amount of cloth needed to make the white stripes.
- Calculate the amount of cloth needed to make the flag.

### Diving Deeper

- Find the length measurements needed to make the flag.
- Estimate or calculate percentages and areas for each color, taking account of the stars.

# CONVERSATION STARTERS FOR #1

*What do you notice? What do you wonder?*

*I wonder* if the flag is drawn to scale?
Yes, it is.

*I wonder* if I have enough information to solve the problem?
You may need to make some measurements.

*I wonder* how many (and which) measurements I need to make?

*I wonder* if I need to know the area of the flag in order to calculate the percentages?

*I notice* that there is one more red stripe than white.

*I notice* (by measuring) that the top stripe is 60% as long as the top of the flag.

# SOLUTIONS FOR #1

*Red*: about 41.5%        *White*: about 36.9%

*Area of red*: 19.8 sq ft        *Area of white*: 17.6 sq ft        *Total area*: $47\frac{2}{3}$ sq ft.

A helpful picture (measurements refer to the original drawing):

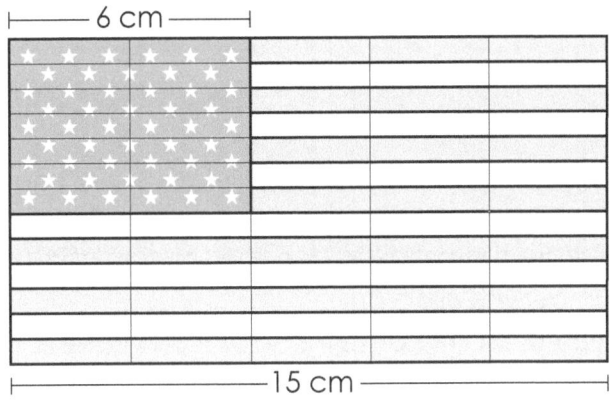

*Note*: Strategies depend on noticing that the top red stripe is $\frac{3}{5}$ of the length of the flag.

As you read the strategies, see if you can understand what students were thinking!

*Some strategies for finding the percentage (red)*:

» $13 \cdot 5 = 65$, $100 \div 65 \approx 1.538$, $1.538 \cdot 27 \approx 41.5$

» $13 \cdot 5 = 65$, $\frac{27}{65} = 27 \div 65 \approx 0.415$, $0.415 \cdot 100 = 41.5$

» $2.2 \div 3 = 0.7\overline{3}$, $0.7\overline{3} \cdot 27 = 19.8$, $0.7\overline{3} \cdot 65 = 47.\overline{6}$, $19.8 \div 47.\overline{6} \approx 41.5\%$

*Some strategies for finding the area (red)*:

» $2.2 \div 3 = 0.7\overline{3}$ ft², $0.7\overline{3} \cdot 27 = 19.8$ ft²
» $2.2 \cdot 9 = 19.8$ ft²
» $3 \div 27 = 0.\overline{1}$, $0.\overline{1} \cdot n = 2.2$, $n = 2.2 \div 0.\overline{1} = 19.8$ ft²
» $47.\overline{6} \cdot 0.415 \approx 19.8$ ft² (using calculations for the percentage of red)

# Problem #2

The Spuddles Potato Chip Company had sales of $2.4 million this year. They have set an ambitious goal to increase sales by 12.5% next year.

## Directions

- Determine the earnings needed to reach the Spuddles company's sales goal. Explain your thinking strategy.
- Illustrate the problem and the solution using a double number line diagram.
- If you used a two-step calculation process, try to discover a one-step method, too.

## Testing the Waters

Solve Problem #2 for $2 million and a 15% sales increase.

# CONVERSATION STARTERS FOR #2

*What do you notice? What do you wonder?*

*Note*: Some students may prefer to write the earnings as $2,400,000.

*I notice* that the answer will be greater than $2.4 million.

*I notice* that the sales goal will be more than 100% of what they made last year.

*I wonder* when it makes sense to talk about percentages that are greater than 100%?

*I notice* that 12.5% is half of 25%.


To make a double number line diagram, let one number line represent sales, and let the other stand for the percentage. Place one number line above the other so that each sales value matches its percentage. (See the Solutions for #1.)

*I wonder* what sales value I should match with 100%?

*I wonder* how much detail I should show in my double number line?
> One suggestion: Divide the percent number line into four equal parts. (Why?)

*I notice* that my double number line is like a table that allows me to see the numbers between the rows or columns!

*I notice* that my double number line shows *equivalent ratios* (0.6 million : 25%, 1.2 million : 50%, etc.) that can be simplified to $24,000 for every 1%.
> In other words, the set of ratios forms a *proportional relationship*. (See Explorations 2 and 3.)

# SOLUTIONS FOR #2

The Spuddles company must earn 2.7 million dollars ($2,700,000) to reach its goal.

*Strategy 1*: Multiply 2.4 million by 12.5% (0.125 or $\frac{1}{8}$) to find the additional sales needed. Add this to the previous year's sales:

$$2,400,000 \cdot 0.125 = 300,000 \qquad 2,400,000 + 300,000 = 2,700,000$$

*Strategy 2*: 2.4 million represents 100%, so the goal is to reach 112.5% of this amount. Divide by 2.4 million by 100 to find 1% of current sales. Then multiply by 112.5:

$$\frac{2,400,000}{100} \cdot 112.5 = 24,000 \cdot 112.5 = 2,700,000$$

*A double number line diagram*:

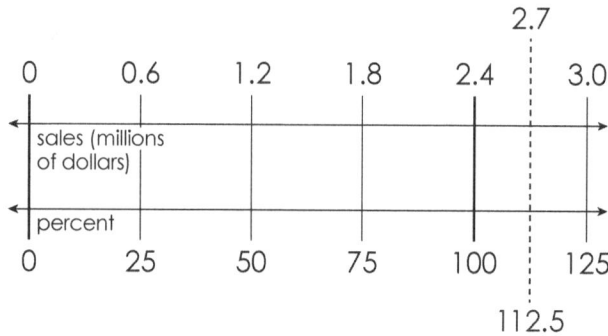

*A one-step strategy*: Multiply 2.4 million by 1.125! 1.125 represents 12.5% more than the whole, or 112.5%. The double number line may help you visualize this. By the way, this strategy is based on the *distributive property*:

$$2,400,000 \cdot (1 + 0.125) = (2,400,000 \cdot 1) + (2,400,000 \cdot 0.125)$$

# Problem #3

The Riverview Middle School Parent-Teacher Organization earned $7100 in its fundraising activities last year. This year, it earned $6600.

## *Directions*

- Create a double number line diagram. Use it to estimate the percent decrease in earnings.
- Develop and explain at least one strategy to calculate the percent decrease.

### *Testing the Waters*

Solve Problem #3 for earnings of $8000 that decrease to $7600.

# CONVERSATION STARTERS FOR #3

*What do you notice? What do you wonder?*

*I wonder* what the "whole" is in this problem?

> Which dollar amount is decreasing? Which quantity are you taking the percentage of?

*I notice* that $500 is less than 10% of $7100.

*I wonder* how carefully I should draw the number lines?

> Remember that you are using them to estimate something.

*I wonder* how to mark the scales on my number lines?

> One possibility: Divide the percent line into 10 equal parts and show the matching intervals of $710 on the earnings line. See the Solutions for #3 for another option.

*I notice* that my double number line shows *equivalent ratios*. They can all be simplified to $71 for every 1%.

> It is easier to see this if you show the dollar amounts for 10%, 20%, etc. As in Problem #2, this means that the relationship is *proportional*.

## After Students Have Solved the Problem

*I wonder* if it matters whether I write 7.0% or 7%?

> It depends on how much precision you want to show. Writing 7.0% shows that you calculated the answer to the nearest tenth of percent instead of the nearest whole percent.

# SOLUTIONS FOR #3

The decrease in earnings is about 7.0%.

*A double number line diagram:*

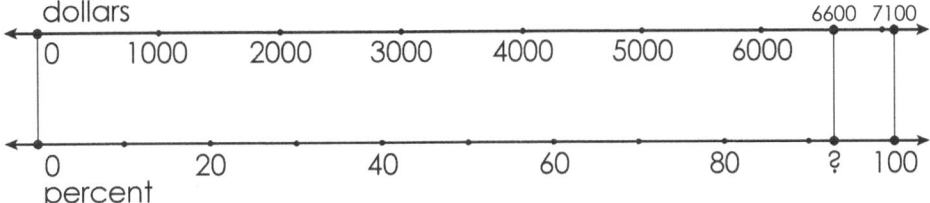

Students' diagrams do not have to be quite this precise. The diagram shows the "whole" ($7100) matched with 100%. The percent decrease from $7100 to $6600 appears to be between 5% and 10%. This makes sense, because 10% of $7100 is $710, and the decrease is only $500.

*Strategy 1*: The rate for this situation is $71 for every 1%. Therefore, divide the number of dollars (6600) by 71 to find the percentage.

$$6600 \div 71 \approx 93$$

This is the percent *of* $7100. The percent *decrease* is about $100\% - 93\% = 7\%$.

*Strategy 2*: Find the difference in earnings, and take it as a percentage of $7100:

$$7100 - 6600 = 500 \qquad 500 \div 7100 = 5 \div 71 \approx 0.070 = 7.0\%$$

*Strategy 3*: Take 6600 as a percentage of 7100. Determine how far this is from 100%.

$$(6600 \div 7100) \cdot 100 \approx 0.930 \cdot 100 = 93.0$$

This is a reduction of about 7.0% from 100%.

# CLASSROOM VIGNETTE FOR #3

Ms. Rodriguez's students have just completed Problem #3. She wants them to recognize that the percentage relationships are a special kind of *proportional relationship*. She reminds them that in a proportional relationship, all of the ratios between inputs and outputs are equivalent. She asks students to make a table.

The class takes a few minutes to decide whether the input should be dollars or percent. They decide to make percent the input, because they think people usually want to know the how many dollars there are for a certain percentage. But they agree that it could be done the other way.

When students are finished making their tables, Ms. Rodriguez chooses one person's table to show on the board (one that makes it easy to see the patterns).

| Percent | 0 | 10 | 20 | 30 | 40 | 50 | 60 | 70 | 80 | 90 | 100 |
|---|---|---|---|---|---|---|---|---|---|---|---|
| Dollars | 0 | 710 | 1420 | 2130 | 2840 | 3550 | 4260 | 4970 | 5680 | 6390 | 7100 |

**Ms. Rodriguez:** How does this table compare to the double number lines you made before?

**Kifah:** It is like the double number lines, because it shows how the dollars and the percents match up, but it doesn't show numbers in between like the number lines do.

**Ms. Rodriguez:** What patterns do you see in the table?

**Jake:** The dollars go up by the same amount each time.

**Ms. Rodriguez:** How much do they go up?

**Javier:** 710 dollars. *(Ms. Rodriguez verifies the response with the class.)*

**Ms. Rodriguez:** Anything else?

**Kifah:** The percent times 71 always equals the number of dollars.

**Ms. Rodriguez:** Could we write that as a formula? Let's use $P$ for percent and $D$ for dollars.

**Ellen:** $D = 71 \cdot P$ *(Ms. Rodriguez writes the formula on the board.)*

**Ms. Rodriguez:** Which variable looks like the input in Ellen's formula?

**Ellen:** I made $P$ the input like we said before.

**Ms. Rodriguez:** Do you agree with Ellen that $P$ is the input in the formula?

**Jake:** Yes, because in her formula, you are doing something to $P$ to figure out $D$.

**Ms. Rodriguez:** Let's look at the number 71 in the formula. Was it helpful to any of you when you solved the problem? *(Many students nod.)* How did it help?

**Kifah:** That's what 1% was. So I divided the dollars by 71 to get the percentage.

**Ms. Rodriguez:** Remember that in many problems we did before, the number we multiplied by to get the output was the same as the amount the output went up by each time. Does that happen this time?

**Kifah:** No, because we multiply each percentage by 71, but the dollar amount goes up by 710 each time.

**Javier:** I agree with Kifah, but I see why this happened. The percents are going up by 10. If the percents went up by one, then the dollars would go up by 71.

**Kifah:** Oh, yeah! That's like what I was doing when I said every 1% was $71!

**Ms. Rodriguez:** So if we ask how much the dollars increase when the percent goes up by 1, then the two numbers are the same? *(The class agrees.)* We have a name for this number. Do you remember it? *(Students seem to have ideas, but no one raises a hand.)* It is also output that goes with an input of 1.

*(The class continues the discussion, reviewing the phrase* unit rate. *They observe that the rate for this situation is $71 per each 1%. Ms. Rodriguez points to examples of students who included the pair 1, 71 in their tables. The class recognizes that this is a proportional relationship because every input/output pair simplifies to this same rate.)*

**Ms. Rodriguez:** Now let's make a graph of the situation. *(Students begin working. Some of them are unsure which variable to put on the horizontal axis. She reminds them that the input goes there. Others need a little help drawing scales on their axes.)*

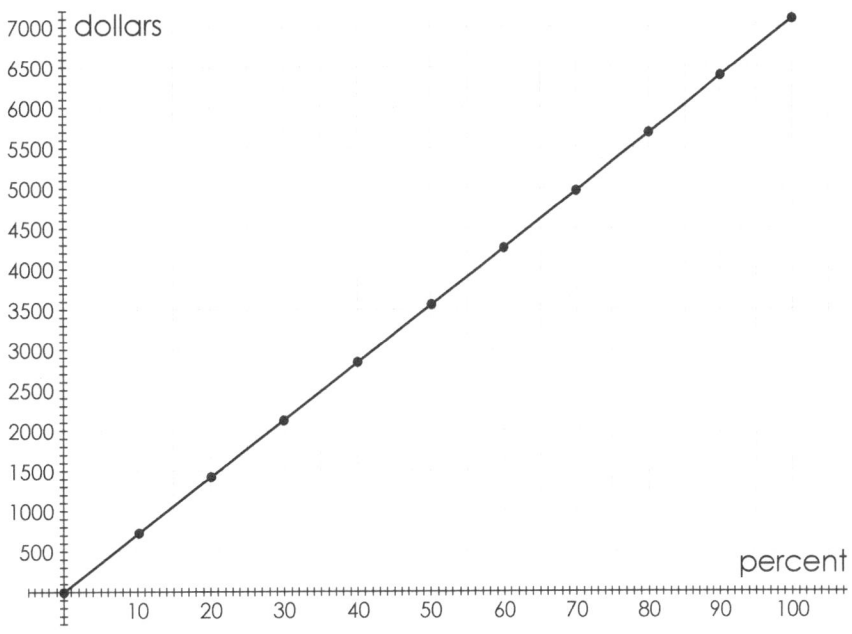

**Ms. Rodriguez:** *(Sharing a graph with the class.)* Do you remember what the graphs of our other proportional relationships looked like?

*(The class recalls that they were straight lines that began at (0, 0). They notice that this graph does the same. They discuss the steepness of the line and how it shows the rate of increase, $71 for each percent.*

*Ms. Rodriguez asks the class to use the table to write equations that show two equal fractions. The students produce many equations, including:*

$$\frac{710}{10} = \frac{1420}{20} \qquad \frac{30}{2130} = \frac{70}{4970} \qquad \frac{40}{2840} = \frac{50}{3550}$$

*They notice that some of the fractions equal the unit rate (71) or its reciprocal. They also see that some fractions compare inputs to outputs, while others compare inputs to inputs and outputs to outputs.)*

**Ms. Rodriguez:** Can you write an equation like this that you could have used to solve Problem #3? *(No one answers.)* Talk it over with your partner for a few minutes.

*(As she watches students work, she notices a few approaches. She chooses the most common one to begin the discussion.)*

**Ms. Rodriguez:** Jake, I saw that you and Javier had something interesting. Would you share it with us, please?

**Jake:** Okay. We got $\dfrac{6600}{7100} = \dfrac{P}{100}$ .

**Ms. Rodriguez:** I see that you used a variable for the number you didn't know. Javier, can you tell us how you and Jake thought about this equation?

**Javier:** Well, we saw that we could compare outputs to outputs like we did a minute ago. We did it that way, because then $P$ out of 100 made sense to be the percent.

**Ms. Rodriguez:** Does anyone have questions or something to add?

**Kifah:** That's not what I got, but it makes sense. You're changing 6600 out of 7100 into some number of parts out of 100, and that's a percentage.

**Ms. Rodriguez:** Is $P$ the answer Problem #3?

**Jake:** No, it's just the percent. Then we have to see how far it is below 100%.

*(Ms. Rodriguez brings in other students' examples and the class talks about them. She summarizes the entire discussion and suggests that students continue to look for unit rates and proportional relationships in the future when they solve percentage problems.)*

# STAGE 2

The first two problems in Stage 2 involve taking percentages of unknown wholes. These kinds of problems often show up in algebra courses, but students learn a lot by solving them with prealgebra knowledge!

The next two problems involve taking percentages of percentages. All sorts of interesting things happen in these situations, because the "whole" changes in the middle of the problem! Students may discover that both of the number lines in their diagrams have units of "percent." They will also wonder whether it affects the solution if they change the starting value in the problem (e.g., the size of the flyer in Problem #6 or the cost of the item in Problem #7).

## *What Students Should Know*

- » Convert between fractions, decimals, and percentages.
- » Find percentages and calculate the percent of a number.

## *What Students Will Learn*

- » Solve percent increase and decrease problems in which the whole is unknown.
- » Use double number line diagrams to represent and understand percentages.
- » Solve percentage problems in which the "whole" changes.
- » Explain patterns in different percentage calculation strategies.
- » Explore percentage situations in which the starting value does not affect the answer.

# Problem #4

You paid $43.42 for a chemistry set, including state sales tax of 6.5%.

## *Directions*

- Estimate the cost of the chemistry set before tax. Create a double number line diagram based on your estimate.
- Use your diagram to determine whether the cost before tax is greater or less than your estimate. Explain your thinking.
- Create a double number line diagram to represent the actual situation.
- Explain at least one strategy to calculate the cost before tax.

## Testing the Waters

Solve Problem #4 if total cost is $42.00 and the state sales tax is 5%.

# CONVERSATION STARTERS FOR #4

*What do you notice? What do you wonder?*

*I notice* that the answer will be less than $43.42 (but not much less).

*I wonder* what the "whole" is in this problem?
> What is being taxed? Which quantity are you taking the percentage of?

*I notice* that the total cost will be more than 100% of the cost before tax.

*I notice* that it does not make sense to take 6.5% of $43.42.

*I notice* that I don't know the dollar amount that I am taking the percentage of!

*I wonder* how I can take a percentage of something that I don't know?
> Try making guesses! Or use a variable to organize your thinking. Does your double number line diagram give you any ideas?

*I wonder* how I can fit the important details on my number line if I start at 0%?
> Maybe you don't have to begin at 0%! Where *is* a good place to start?

*I wonder* how I can know if my answer is correct?
> Test it! Calculate the tax and add it on. If the answer is wrong, you'll know how to revise it!

*I wonder* if it's possible to calculate the answer in one step?
> Look back at the final part of Problem #2. Can you adapt it to this situation?

*I wonder* if it is possible to use a variable to analyze this situation?

# SOLUTIONS FOR #4

The cost of the chemistry set before tax is $40.77.

*An estimate and its double number line diagram*: The cost before tax should probably be about $40.

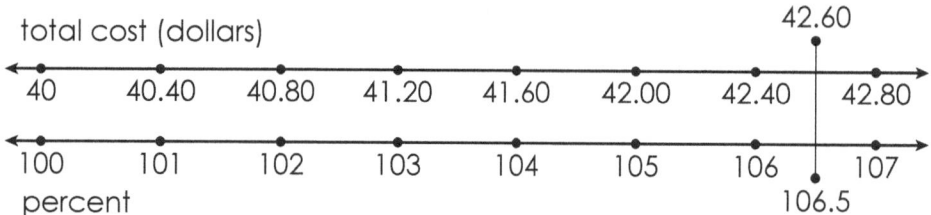

The double number line shows that 106.5% of $40 is $42.60, which means that the estimate was too small. You could adjust the estimate until you find a solution.

*A double number line diagram for the actual situation*:

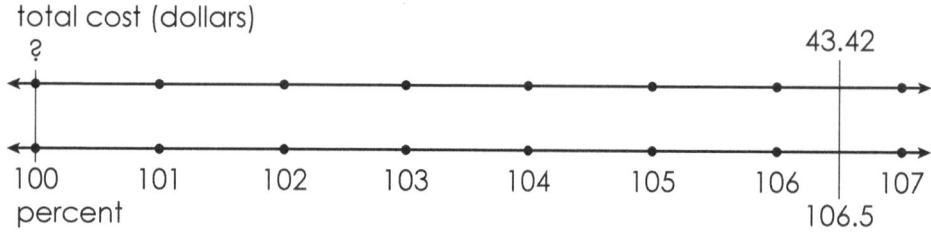

The "whole" (cost without tax) is the value on the cost line that matches with 100%.

*Calculation 1*: Each interval between the dots represents 1%. Because there are 106.5 intervals between $0 and $43.42, each interval represents $43.42 \div 106.5 \approx 0.4077$ dollars. There are 100 of these intervals between 0% and 100%, so the cost without tax is approximately $0.4077 \cdot 100 = 40.77$ dollars.

*Calculation 2*: Divide the total cost (with tax) by 106.5% or 1.065:

$$43.42 \div 1.065 \approx 40.77 \text{ dollars}$$

# Problem #5

The town of Lonesome had a population of 45 people 10 years ago, which is about 67% more people than it has now.

*Directions*

- Guess the current population. Use your estimate to create a double number line diagram.
- Use your diagram to determine whether the current population is greater or less than your guess. Explain your thinking.
- Create a double number line diagram to represent the actual situation.
- Develop and explain at least one strategy to calculate the current population.

## *Diving Deeper*

Write algebraic equations to represent Problems #2 through #5.

## *Testing the Waters*

Solve Problem #5 using 60 people and 50%.

# CONVERSATION STARTERS FOR #5

*What do you notice? What do you wonder?*

*I notice* that Lonesome is a good name for this town!

*I notice* that 67% is close to $\frac{2}{3}$.

*I wonder* what the "whole" is in this problem?

What are you comparing the known population to? Which quantity are you taking the percentage of?

*I notice* I am taking the percentage of an unknown amount (just like in Problem #4).

*I notice* that showing 0% on my number lines makes it hard to show some important details.

*I notice* that something feels "backward" when I draw my number line.

The current population is probably on the left side of your line, while the past population is on the right!

*I wonder* why I get a fractional number of people for an answer?

Notice that the original problem says that the percentage is *about* 67%. It's the percentage that isn't a whole number, not the number of people!

# SOLUTIONS FOR #5

The town of Lonesome has a population of 27 people.

*A guess and a double number line diagram*: The population is probably a little less than 30, because 67% is about two-thirds, and if you increase 30 by two-thirds of itself, you get $30 + 20 = 50$, which is too high.

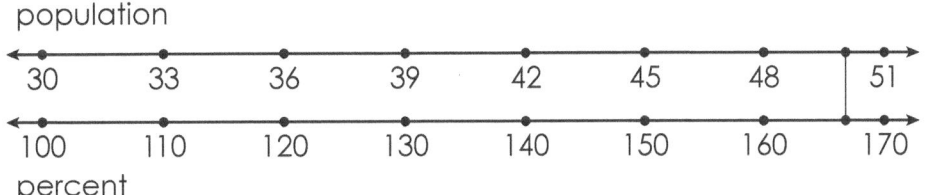

population

| 30 | 33 | 36 | 39 | 42 | 45 | 48 | 51 |

| 100 | 110 | 120 | 130 | 140 | 150 | 160 | 170 |

percent

The double number line also shows that 167% of 30 is about 50, so the estimate was too large. You could keep adjusting the estimate until you find a solution!

*A double number line diagram for the actual situation*:

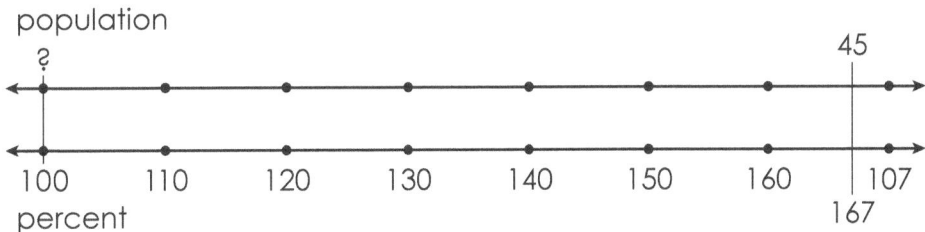

population

The "whole" (the current population) is the value on the population line that matches with 100%.

*Calculation strategy 1*: Each interval between the dots represents 10%. Because there are 16.7 intervals between 0 and 45 people, each interval stands for $45 \div 16.7 \approx 2.7$ people. There are 10 intervals between 0% and 100%, so the current population is $2.7 \cdot 10 = 27$ people.

*Calculation strategy 2*: Divide the original population by 167% or 1.67:

$$45 \div 1.67 \approx 27 \text{ people}$$

# Problem #6

You are creating a flyer for a band concert. You used a copier to enlarge it to 160% of its original size. You have lost the original flyer.

## Directions

- Guess the copier setting that will return the flyer to its original size. Test your guess.
- Create a double number line diagram to represent the problem.
- Show a strategy to calculate the answer without having to guess and test numbers.

## Diving Deeper

If the copier allows you to enter only whole number percentages:

- In Problem #6, it possible to make a copy exactly the same size as the original?

- Is it ever possible to make a copy that is one-third the size of an original? How close can you get if you are able to make two copies?

## Testing the Waters

Solve Problem #6 using 200%, 150%, or 125%.

# CONVERSATION STARTERS FOR #6

*What do you notice? What do you wonder?*

*I wonder* if the 160% refers to the side lengths or the area of the poster?

    It refers to the side lengths.

*I notice* that my first guess of 40% doesn't make sense.

    The answer must be greater than 50%, because 50% of 160% is 80%, which is less than the size of the original flyer.

*I wonder* if the answer depends on the size of the original flyer?

    Try experimenting with different sizes. (It does not!)

*I wonder* why it's so hard to determine the "whole" in this problem?

    It changes! At first, the whole is the size of the original flyer. Later, it becomes the size of the enlarged flyer.

*I notice* that both of my number lines are labeled "percent."

    This happens because the problem involves finding percentages of percentages.

## After Students Have Solved the Problem

*I wonder* what the connection is between 160% and my answer of 62.5%?

    They are reciprocals! You can test this by multiplying: $160\% \cdot 62.5\% = 1.6 \cdot 0.625 = 1$

# SOLUTIONS FOR #6

Set the copier to 62.5% (or as close to this as possible).

*Making a guess*: Students often guess 40%. Let's test it! Suppose the flyer is 10 inches tall. The 160% enlargement will make it 16 inches tall. 40% of 16 inches is 6.4 inches. The guess is too small because 6.4 inches is less than the original 10 inches.

*A double number line diagram:*

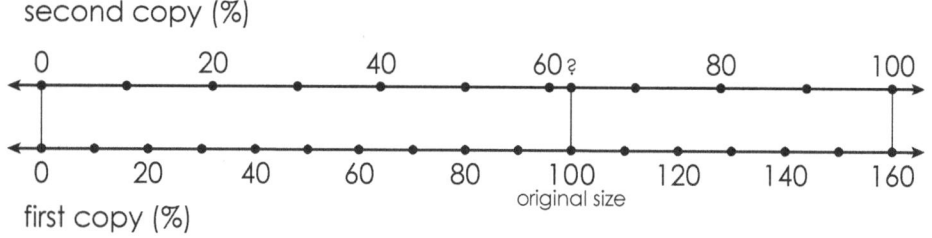

*Strategy 1*: 160% of the original size is the new whole (the new 100%). We are looking for the percentage on the top number line that matches with 100% on the bottom line. Each 10% interval on the bottom line is $100 \div 16 = 25 \div 4 = 6.25$ (percent) on the top line. 10 of these intervals is $6.25 \cdot 10 = 62.5$ (percent) on the top line.

*Strategy 2*: The double number line shows that the question is "What percent of 160 is 100?" $\dfrac{100}{160} = \dfrac{5}{8} = 62.5\%$. (In decimal form, $1 \div 1.6 = 0.625$.)

*Strategy 3*: The enlarged copy is $\dfrac{8}{5}$ of the original size. The new one must be $\dfrac{5}{8}$ (62.5%) of this in order to return it to its original size. The copier settings are reciprocals!

# Problem #7

There is a 20% discount on all items at your favorite store. In addition, you have a coupon for 30% off the discounted price.

## *Directions*

- Create and use a double number line diagram to estimate the percentage off the original price you will receive.
- Find the solution using the double number line or another strategy.
- Explain what happens if you take the discounts in reverse order.

### *Diving Deeper*

Create an algebraic expression for the percentage off the original price.

### *Testing the Waters*

Solve Problem #7 if both numbers are 50%.

# CONVERSATION STARTERS FOR #7

*What do you notice? What do you wonder?*

*I notice* that the discount is not 50%.

> It is not 50%, because the 30% is not taken off the original price.

*I notice* that the discount will be less than 50%.

> The second reduction is a percentage of a smaller amount.

*I notice* that $100 is a convenient price to work with!

*I wonder* if the answer depends on how much I spend?

> Experiment with different prices! (The percentage discount is the same regardless of the original price.)

*I notice* that the "whole" changes in this problem (just like in Problem #6).

> The original whole is the price of the item before any discounts are taken. The new whole is the price after the 20% discount.

*I notice* that 20% *off* the original price is 80% *of* the original price.

*I notice* that I can take percentages of percentages rather than a percentage of a price.

*I notice* that I can use a variable to represent the price.

# SOLUTIONS FOR #7

You will receive 44% off the original price.

*A double number line diagram*:

second discount (%)

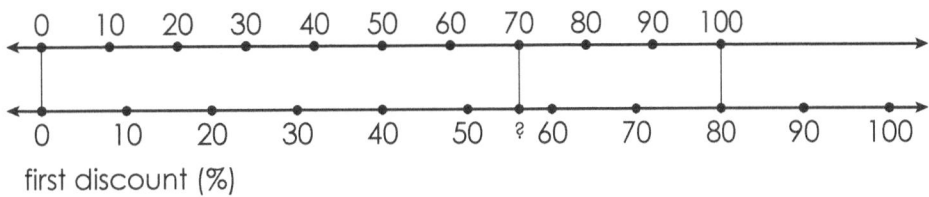

first discount (%)

The answer to the problem is the amount (percentage) below 100 of the "?" on the bottom line.

*Strategy 1*: Choose a price. If you choose $100, then 1% represents 1 dollar! 20% off of $100 dollars is $80. 30% off of $80 is $80 − $24 = $56, which is $44 less than $100 or 44% off the original price. If you choose other values for the price, you will still get an answer of 44%! (Try it!)

*Strategy 2*: Each 10% on the top line of the diagram represents $80 \div 10 = 8$ (percent) on the bottom line. Thus, 70% on the top line is $8 \cdot 3 = 24$ (percent) below 80% on the bottom line. $80 - 24 = 56$, and 56% is 44% less than 100%.

*Strategy 3*: Let $P$ stand for the original price. 20% off of this price is 80% of the price, which is $0.8 \cdot P$. 30% off of this price is 70% of it, which is $0.7 \cdot (0.8 \cdot P) = (0.7 \cdot 0.8) \cdot P$, or $0.56 \cdot P$. This expression represents 56% of $P$ or 44% *off* the original price.

If you take the percentages in reverse order, the result will be the same, because:

$$0.7 \cdot 0.8 = 0.8 \cdot 0.7 \text{ (the } commutative \text{ property of multiplication)}$$

# STAGE 3

Problem #8 shows a more complex situation involving percentages of percentages. Your students may find that double lines are less helpful here. Some of them may wonder if a triple number line would work or if they could use two double number lines. Encourage them to explore these possibilities.

Problem #9 is mainly for fun. It illustrates a different type of percentage pattern. It is also an example of a seemingly complex problem that can be made much simpler by using some flexibility and ingenuity. The answer is surprising!

## What Students Should Know

» Calculate percentages and find the percent of a number.

## What Students Will Learn

» Solve complex and challenging problems involving percentages.

# Problem #8

Tabby and Abdullah are running for class president. 80% of the votes have been counted. Of these, Tabby has received 40%, and Abdullah has received 60%.

## Directions

- Determine if it is possible for Tabby to win the election. Explain your reasoning.
- If it is possible, find the percentage of the remaining votes must Tabby must receive.
- If it is not possible, calculate the greatest percentage of the total vote Tabby can receive.

# CONVERSATION STARTERS FOR #8

*What do you notice? What do you wonder?*

*I notice* that 60% − 40% = 20% and that 20% of the vote remains to be cast.

*I wonder* if this means that Tabby needs all of the remaining votes just to tie?

*I wonder* if the answer depends on the total number of voters?
> Experiment with different numbers of voters! (Just like Problems #6 and #7, the answer does not depend on the number in the original whole.)

*I wonder* how many different "wholes" are involved in this problem?
> You are taking percentages of three different quantities: the total available votes, the votes that have been cast, and the votes remaining to be cast.

*I wonder* if I could make a triple number line diagram for this problem?
> Try it! It might work. Or maybe you would rather have two double number lines. You may even decide that number lines are not very helpful in this case.

# SOLUTIONS FOR #8

It is possible for Tabby to win. She needs to receive just more than 90% of the remaining votes.

*Strategy 1*: Suppose there are 150 voters.

| Number of votes cast | 80% of 150 | 120 votes |
|---|---|---|
| Number of votes for Tabby | 40% of 120 | 48 votes |
| Total votes Tabby needs (to tie) | 50% of 150 | 75 votes |
| Additional votes Tabby needs (to tie) | 75 − 48 | 27 votes |
| Votes remaining to be cast | 150 − 120 | 30 votes |
| Percent of remaining votes needed (to tie) | 27 of 30 | 90% |

The answer will be 90% to tie regardless of the total number of voters. I used 150 voters for the sake of variety, although it would have been easier to work with 100 voters. Try it!

*Strategy 2*: Tabby has received 40% of 80% of the vote, which is $0.4 \cdot 0.8 = 0.32$ or 32% of the total vote so far. She needs another 18% of all votes in order to get 50% of the vote, but only 20% of the votes remain to be counted. 18 is 90% of 20, so she must receive more than 90% of the remaining vote.

# Problem #9

A lemonade mixture contains 99% water by volume. It is left outside in the sun and some of the water evaporates, resulting in a mixture that is 98% water.

## Directions

- Represent the situation with a diagram, table, and/ or graph.
- Determine the percentage of the original mixture that evaporated.
- Describe your strategy.

# CONVERSATION STARTERS FOR #9

*What do you notice? What do you wonder?*

*I wonder* what "by volume" means?
>The amount of lemonade could be measured by either weight or volume.

*I wonder* what is in the "nonwater" part of the mixture?
>It probably contains lemon and sugar. (However, the water in the lemon is considered to be part of the 99% water in the mixture.)

*I notice* that 99% and 98% represent percentages of the entire mixture.

*I notice* a pattern in my fractions comparing the water to the whole mixture.
>The denominator is always 1 greater than the numerator, even after water evaporates.

*I wonder* how it affects the percentage of water in the mixture when 1% of the mixture evaporates?

*I wonder* if each 1% decrease in the original mixture will affect the percentage of water in the mixture by the same amount?

*I notice* that it helps to focus on the part of the mixture that is not water!

# SOLUTIONS FOR #9

50% of the original mixture evaporated!

*A table*:

| Percent of Mixture Evaporated | 0 | 1 | 2 | 3 | 4 | 5 |
|---|---|---|---|---|---|---|
| Fraction of Water in the Mixture | $\frac{99}{100}$ | $\frac{98}{99}$ | $\frac{97}{98}$ | $\frac{96}{97}$ | $\frac{95}{96}$ | $\frac{94}{95}$ |
| Percentage of Water in the Mixture | 99.00 | 98.99 | 98.98 | 98.97 | 98.96 | 98.95 |

The water in the mixture appears to be decreasing by about 0.01% for every 1% increase in the mixture evaporated. At this rate, the percentage of water would not reach 98% until the water is gone, which does not make sense!

*Another table*:

| Percent of Mixture Evaporated | 10 | 20 | 30 | 40 | 50 |
|---|---|---|---|---|---|
| Fraction of Water in the Mixture | $\frac{89}{90}$ | $\frac{79}{80}$ | $\frac{69}{70}$ | $\frac{59}{60}$ | $\frac{49}{50}$ |
| Percentage of Water in the Mixture | 98.89 | 98.75 | 98.57 | 98.33 | 98.00 |

This table shows that the percentage of water is gradually decreasing more quickly as the total amount of lemonade becomes smaller. The answer (50%) appears in the final column.

*A different thinking strategy*: In order for the percentage of "nonwater" to double from 1% to 2%, the total volume of the lemonade mixture must become half (50%) of what it was.

*I wonder* what will happen to the percentage of water in the lemonade as it continues to evaporate? What will this look like if I graph it?

# ALGEBRA CONNECTIONS

Prealgebra students may try developing algebraic equations for these problems by organizing their work from guess and test processes. Using Problem #4 as an example (with parentheses to represent multiplication):

| Cost of the set (guess) | Tax | Total |
|:---:|:---:|:---:|
| 40.00 | $0.065(40.00)$ | $40.00 + 0.065(40.00) = 42.60$ |
| 41.00 | $0.065(41.00)$ | $41.00 + 0.065(41.00) \approx 43.67$ |
| 40.50 | $0.065(40.50)$ | $40.50 + 0.065(40.50) \approx 43.13$ |
| $x$ | $0.065x$ | $x + 0.065x = 43.42$ |

The number they are guessing is the variable, and the final equation is in the "Total" column! For this process to succeed, students must write down their calculation process as they work. Those who know how to solve linear equations may continue.

$$x + 0.065x = 43.42$$
$$1.065x = 43.42$$
$$x \approx 40.77$$

Algebra students who understand connections between percentages and proportional relationships may also solve some of the percentage problems by writing and solving algebraic proportions. For example, Problem #4 may be solved using the equations:

$$\frac{100}{106.5} = \frac{x}{43.42} \quad \text{or} \quad \frac{x}{100} = \frac{43.42}{106.5}$$

Notice how these two equations connect to the double number line shown in the Solutions.

# Exploration 5

## Scaling a Tower

In this exploration, students resolve a friendly disagreement between Derek and Beth, who have made different predictions about the height of a transmission tower near their home. As they create a mathematical model (a scale drawing in this case) and use it to solve the problem, they learn more about *proportional relationships* and begin to explore *similar* shapes.

The solution pages show a variety of strategies. However, you may notice that the familiar "cross multiplication" method is not one of them. Why not simply show students the easiest and most efficient process right away? To me, the main difference is between teaching them to follow rules and helping them learn to think for themselves. They learn more when they solve these problems using their own ideas, because they must pay attention to the relationships in their drawings. They also develop the flexibility to solve problems in more than one way.

Observing students as they devise problem-solving strategies enables you to recognize and correct students' misconceptions and to learn from their insights. When the shortcuts eventually come (as cross multiplication does in Exploration #6), students appreciate and understand them more for having wrestled with the ideas themselves.

# STAGE 1

Before you hand Problem #1 out to students, try gathering their ideas first! Describe the situation in the first paragraph, and ask them how they might determine the height. What measurements could they make? How would they use them? What additional factors should they consider? After this discussion you can follow up on their ideas, modify the directions, or simply launch into the problem as written.

A couple of tips before students begin work:

» Discuss the *angle of elevation*: the angle between a horizontal line and the line between you and the top of the tower. (The Solutions page shows what this looks like.)

» Ensure that you get drawings of various sizes. Students will compare them later.

## What You Will Need

» Rulers and protractors (and possibly compasses).

» Large pieces of paper or poster board (for those who make larger drawings).

» Scientific calculators that have a "tan" key for the *tangent* function. (This is needed for Stage 3 only.)

## What Students Should Know

» Measure and draw lengths (with customary and metric units) and angles.

» Understand the meaning of a *proportional relationship* (see Exploration #2).

## What Students Will Learn

» Create and use scale drawings to measure real-world quantities.

» Observe connections between scale drawings and proportional relationships.

» Begin to explore (or enrich their understanding of) similar shapes.

# Problem #1

There is a transmission tower in a flat fenced field near Beth and Derek's house. Derek thinks the tower is more than 50 meters tall. Beth disagrees. To settle the issue, they design a plan to measure the height of the tower.

Using a map, they determine that the field is about 88 meters long. Standing on opposite sides of the field with the tower directly between them, Beth measures the angle of elevation to the top of the tower to be 38°. Derek makes the same measurement from his side and gets 61°.

## Directions

- Make a scale drawing of the situation.
- Use your drawing to determine the approximate height of the tower.
- Show your calculations and explain your thinking.

### Diving Deeper

Find a tall object. Make length and angle measurements. Create a scale drawing, and use it to estimate the height of the object.

### Testing the Waters

How high would the tower be if each person measured an angle of elevation of 45°?

# CONVERSATION STARTERS FOR #1

*What do you notice? What do you wonder?*

*I wonder* what the angle of elevation would be if you were standing right under the tower?

*I wonder* why Derek and Beth got different angle measurements?

*I wonder* if Beth and Derek should lie down when they measure the angles, so that they are closer to the ground?

*I wonder* if it matters how large I make my drawing?

*I wonder* if I can use sides of squares on graph paper as the units in my drawing?
> Yes. In fact, it may be easier than measuring with a ruler. However, you should think about how accurately you can determine measurements between the grid lines.

*I notice* that the triangle in my drawing is *similar* to a triangle connecting Beth, Derek, and the top of the tower.

*I notice* that the height of the tower is measured at a right angle to the ground.

*I wonder* if I have to guess where to draw the point on the ground below the tower?
> Many students try different points, measuring to see if the height segment makes a 90° angle with the ground. Some may discover creative ways to use a compass to find this point!

*I notice* that my calculations are easier when I use metric units.

*I notice* that millimeters are smaller (more precise) than sixteenths of an inch.

*I wonder* how I should round my answer?
> Use your best judgment. You will explore this question in more detail in the next problem!

*I wonder* what happens when an angle of elevation doubles? Does the height double?

## After Students Have Found Answers

*I wonder* if I can tell who is right when the answer is so close to 50 meters?

*I wonder* how Derek and Beth could find the answer if the field were on a slope?

# SOLUTIONS FOR #1

Most students' solutions should be between about 46 and 51 meters.

*A sample drawing and measurements:*

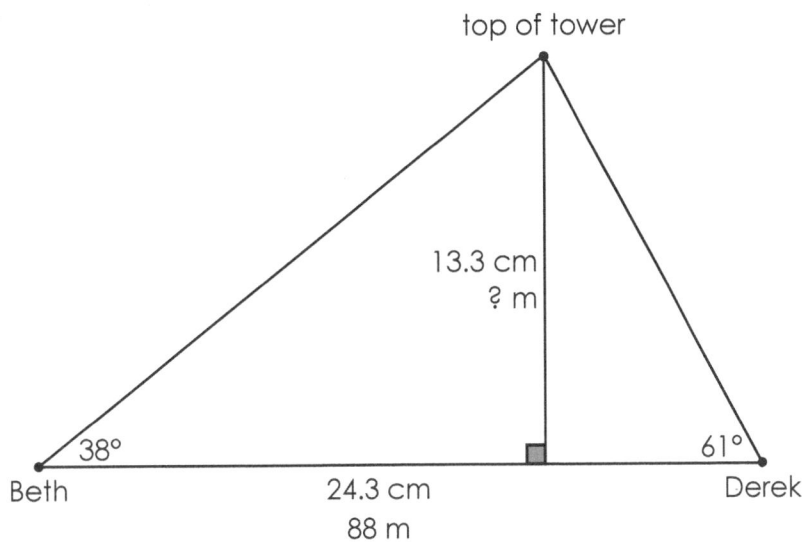

*Calculation strategy 1:* Each 1 cm in this drawing represents $88 \div 24.3 \approx 3.621$ m in the real world. The tower is about $13.3 \cdot 3.621 \approx 48.2$ meters high.

*Calculation strategy 2:* The height of the triangle in the drawing is about $\frac{13.3}{24.3} \approx 54.73\%$ of the base. The top of the tower is about 54.73% of 88 m above the ground:

$$88 \cdot 0.5473 \approx 48.2 \text{ m}$$

*Calculation strategy 3:* The base of the triangle in the drawing is about $24.3 \div 13.3 \approx 1.827$ times as long as the height. 88 m is about 1.827 times as long as the height ($h$) of the tower:

$$1.827 \cdot h \approx 88 \qquad h \approx 88 \div 1.827 \approx 48.2 \text{ m}$$

Do you have students who always like to report their answers with as many decimal places as their calculator shows? Do your textbooks always show solutions as precise numbers? These things may make sense for purely mathematical problems, but students need to learn that solutions are rarely exact when it comes to real-world situations.

The processes of modeling, measuring, and calculating all affect the accuracy (correctness) and precision (exactness) of answers to real-world problems. In Stage 2, students pay close attention to these processes in order to make decisions about what conclusions they can draw and how they should report their answers.

## What Students Should Know

&raquo; Find the mean and median of a set of numbers.

## What Students Will Learn

&raquo; Accuracy refers to correctness, and precision is about exactness.

&raquo; Real-world measurements are never exact.

&raquo; Make informed judgments about the precision of solutions to real-world problems.

&raquo; Use judgments about precision to draw conclusions and decide how to report answers.

# Problem #2

When you use measurements in a calculation, it is important to think about how accurate and precise your answer is likely to be.

## *Directions*

- Name some things that might limit the precision of your answer to Problem #1.
- Determine how your answer is affected if a measurement in your drawing is off by 1 millimeter, 1 sixteenth of an inch, or 1 degree.
- Compare the solutions of students in your class. Find a typical value for the class.
- Revise your answer to Problem #1 if necessary. Explain your thinking.

### *Diving Deeper*

Continue the "Diving Deeper" activity from Stage 1 by using methods from Problem #2 to estimate the precision of your answer. Report your answer as an interval.

# CONVERSATION STARTERS FOR #2

*What do you notice? What do you wonder?*

*I wonder* how precise the measurements in the original problem are?

It depends on how Beth and Derek made their measurements. A practical rule of thumb is to assume that the final digit is correct within about 2 units of the smallest place value. For example, because the field's length is shown as 88 meters, the actual distance may be between about 86 and 90 meters.

*I notice* that rounding numbers in the middle of my calculations changed my answer.

*I wonder* how the size of my drawing affected the exactness of my answer?

*I wonder* what is the best way to combine the class's data?

*I wonder* if there is a good way to graph the class's data?

*I wonder* if I should report my answer as a single number or as an interval?

Which gives a better idea of how close your answer is likely to be to the actual height?

# SOLUTIONS FOR #2

Some things that may affect your answer for the tower's height:

» How precise Beth and Derek's measurements were.

» How carefully you drew your picture and made your measurements.

» The number of decimal places you used in your calculations.

» The size of your drawing (if you could measure perfectly, the size of the drawing would not matter, but larger drawings will give more *precise* answers).

*Effects of measurements on answers*: For the picture in the Solutions for Problem #1, every millimeter represents about 0.36 meters in the real world. If the height measurement is off by 1 mm, the answer will be off by about 0.36 m. If this measurement is off by 2 or 3 mm, the predicted height could easily be off by over a meter in either direction. For smaller pictures, the effect is even larger, because each 1 mm in the drawing represents a greater real-world distance.

To see what happens when you change an angle by 1°, you need to redraw the picture with the new angle. You may notice an even greater effect on the answer. This is important, because it may have been difficult for Beth and Derek to measure these angles accurately.

At this point, the evidence seems to favor Beth's prediction, but there is room for doubt.

*Combining the class data (Option 1)*: Use the mean or median of the class data to find a typical value. If there are *outliers* (numbers far away from most of the other values), it is probably better to use the median.

*Combining the class data (Option 2)*: If you know how to create a *histogram* of the data, use it to make a visual estimate.

*Combining the class data (Option 3)*: If you know how to calculate *mean absolute deviation* or *interquartile range*, use them to estimate an interval in which the answer probably lies.

*Revisions to students' answers*: Based on the class data, students may revise their answers to approximately 48–50 meters. Individual answers may lie outside this interval, but the answer you get from the *combined* data is probably more precise (assuming the numbers in the original problem were measured carefully)!

# STAGE 3

In my experience, Problems #3 and #4 tend to work best as whole-group activities. I often choose not to give students a handout for Problem #3, because if I do, they get impatient to learn about the *tangent*, and it is hard for them to focus on the measurement comparisons at the beginning of the problem. Making these comparisons is at the heart of developing an understanding of what the tangent of an angle really is.

In Problem #4, students learn that it is possible to answer Problem #1 without making a scale drawing! Instead, they apply knowledge from Problem #3 to create and solve a system of algebraic equations. The trial and error process develops number sense as it paves the way to a stronger understanding of algebra concepts for the future.

## What Students Should Know

» Understand the meaning of a *proportional relationship*.
» Understand the meaning of *similar* geometric figures.

## What Students Will Learn

» Explore ratios in similar triangles.
» Write algebraic equations to model a real-world situation.
» Develop a trial and error method to solve a system of algebraic equations.
» Begin to explore the *tangent* relationship in trigonometry.

# Problem #3

Your drawing contains a right triangle with a 38° angle. The side lengths of this triangle are related in predictable ways.

*Directions*

- Compare the vertical and horizontal sides of the right triangle in your drawing.
- Combine the class's results to get a better comparison. Explain what happens and why.
- Use a scientific calculator to find the value of tan 38° (the *tangent* of 38°). Describe what you notice.
- Repeat the process above for the right triangle with a 61° angle.
- What do you think the *tangent* of an angle is? Explain.

## Diving Deeper

Continue the Diving Deeper activity from Stages 1 and 2 by using the *tangent* function to obtain a better result for your estimate.

## Testing the Waters

Explore Problem #3 for a 45° angle.

# CONVERSATION STARTERS FOR #3

*What do you notice? What do you wonder?*

*I wonder* if I should use subtraction or division to compare the side lengths?
    If you are not sure, try it both ways!

*I wonder* if it matters in which order I subtract or divide the side lengths?
    Again, do it both ways. See what happens.

*I wonder* if it matters whether I use inches or centimeters when I divide?

*I notice* that when I subtract the side lengths in both orders, the answers are opposites.

*I notice* that when I divide the side lengths in both orders, the answers are reciprocals.

*I wonder* what the answer means when I divide the lengths?
    Some examples: If the answer is 1.4, then one side is 1.4 times longer than the other. If the answer is 0.37, then one side is 0.37 times as long (37% as long) as the other.

*I notice* that when everyone uses division (in the same order), their answers are almost the same even though their triangles are different sizes.

*I wonder* why this happens?
    What is the relationship between everyone's triangles? (They are all *similar*, which means that they are all enlarged or reduced versions of each other.)

*I notice* that the value of tan 38° is about the same as one of the division comparisons.

*I wonder* if the people who drew larger triangles got better answers (closer to tan 38°)?
    Their answers will often be more precise.

*I wonder* if I can use the *tangent* to get a better answer for the height of the tower?
    You will have a chance to explore this question in Problem #4!

*I wonder* what the *sin* and *cos* keys on the calculator are for?
    They are also about angle relationships in right triangles. They stand for *sine* and *cosine*. Experiment! Maybe you can figure out what they mean.

# SOLUTIONS FOR #3

*Subtraction comparisons*: Students get different answers for triangles of different sizes when they subtract.

*Division comparisons*:
  » longer ÷ shorter: between about 1.2 and 1.4
  » shorter ÷ longer: between about 0.76 and 0.80

Everyone's values for the division comparisons (ratios) are nearly the same. This occurs because their triangles are similar. The comparisons would be exact if it were possible to draw and measure the sides perfectly.

*Combining the class's division comparisons*: Students usually use the mean (average) to combine the measurements.
  » longer ÷ shorter: about 1.28
  » shorter ÷ longer: about 0.78

*Value of the tangent of 38°*: $\tan 38° \approx 0.78129$. The value of tan 38° is about the same as the shorter ÷ longer comparison above.

*Value of the tangent of 61°*: $\tan 61° \approx 1.80405$. Students' division comparisons should be nearly the same as this number to the nearest hundredth (or at least tenth).

*Meaning of the tangent*: The *tangent* of an angle is a division comparison (ratio) between the lengths of the two shorter sides of a right triangle: the side opposite the angle divided by the side adjacent to it. Students may need to explore this side relationship with other triangles in order to to reach this conclusion.

# CLASSROOM VIGNETTE FOR #3

Ms. Rodriguez is not using the handout for Problem #3. She plans to guide her students through the problem so that they can discuss it as they work. She asks the class to take out their drawings and measure the vertical and horizontal sides of the right triangle with the 38° angle, telling them that they will be comparing everyone's measurements later.

**Kifah:** Does it matter if we use inches or centimeters?

**Ms. Rodriguez:** What does everyone think?

*(The class discusses it and decides on centimeters, because they can write them as decimals, which will be easier to compare. While students make the measurements, Ms. Rodriguez observes and assists with individual questions.)*

**Ms. Rodriguez:** Now that you have made your two measurements, I want you to think of a calculation you can do to compare them.

**Jake:** How many decimal places should we show?

**Ms. Rodriguez:** I'm going to suggest four. We can round more later if we want.

*(Some of her students look uncertain, because she has not told them what calculation to do. She reassures them that they can choose a calculation that makes sense to them. When she sees some students waiting for others to finish, she suggests that they find and record a second way to compare the numbers.)*

**Ms. Rodriguez:** Okay. Now, we're going to collect everyone's answers and put them on the board. Just share your answer. Don't tell us how you got it. We're going to figure that part out together. *(She records the answers on the board.)*

| 0.75 | 3.7 | 1.2791 | 1.3426 | 1.6 | 0.7824 | 0.7714 |
| 7.2 | 1.3118 | 0.7801 | 1.2818 | 0.781 | 0.7964 | |

**Ms. Rodriguez:** Do you notice anything?

**Javier:** Some of the answers are really close.

**Ms. Rodriguez:** I see what you mean! Which ones are *not* close? *(Students look briefly.)*

**Kifah:** The short ones all look different from each other.

**Ms. Rodriguez:** That's interesting. What do you mean by the short ones?

**Kifah:** 3.7, 1.6, and 7.2. The numbers that stop at the tenths.

**Ms. Rodriguez:** Maybe we can group the answers that belong together. Take a minute to do that.

*(The class agrees on three groups. Ms. Rodriguez writes them in rows on the board.)*

| 3.7 | 1.6 | 7.2 | | | |
|--------|--------|--------|--------|-------|--------|
| 0.75 | 0.7824 | 0.7714 | 0.7801 | 0.781 | 0.7964 |
| 1.2791 | 1.3426 | 1.3118 | 1.2818 | | |

**Ms. Rodriguez:** Do you think numbers within each group were calculated the same way?

**Ellen:** I don't think the ones in the first row were. They're too different from each other.

**Jake:** I don't agree with Ellen, because they're all tenths. That's probably what you'd get if you add or subtract.

**Ms. Rodriguez:** Say a little more.

**Jake:** Well, I think most people would subtract or divide. But if you divide, you usually get lots of decimals.

**Ms. Rodriguez:** Did everyone subtract or divide? *(Everyone nods.)* Why do you think the answers came out so different for the people who got 3.7, 1.6, and 7.2?

**Kifah:** I think it depends on how big their picture was. People with bigger pictures got bigger answers when they subtracted.

*(The class checks with the people who subtracted and discovers that Kifah is correct.)*

**Ms. Rodriguez:** What about the people who divided? Why are their answers so close?

**Ellen:** Maybe because when the pictures are just different sizes, it sort of balances out.

**Ms. Rodriguez:** *(Waiting a moment)* Can you say any more? *(Ellen looks unsure.)*

**Ms. Rodriguez:** Does anyone else understand what Ellen is thinking?

**Javier:** It's kind of like the ramps we did before. [See Exploration 2.] We used dividing to compare their sides, and it didn't matter how big they were, just how steep.

**Ms. Rodriguez:** Does that fit with what you were thinking, Ellen?

**Ellen:** I wasn't really thinking about the ramps, but it's the same idea.

**Ms. Rodriguez:** If we look at the longest side of our triangles, do you think everyone's has the same steepness? *(Most students say yes. Some compare their pictures.)*

**Ms. Rodriguez:** We will learn later that all of these are called *similar triangles*, because they look the same except that they are larger or smaller. Can you explain why there are two groups of answers for the people that divided?

**Kifah:** Some people did bigger divided by smaller, and the others did smaller divided by bigger.

**Ms. Rodriguez:** Can we tell who did which one?

*(The discussion continues, and Ms. Rodriguez suggests that the students who divided the same way combine their answers. The students decide to use the mean for this. Ms. Rodriguez shows the two means on the board: 0.777 and 1.304.)*

**Ms. Rodriguez:** What do these numbers tell us about how the sides compare?

**Kifah:** The 1.304 means that the long side is about 1.3 times longer.

**Ms. Rodriguez:** What about 0.777? *(The students are unsure.)* What if it were 0.75?

**Jake:** Oh, I see! The short side would be $\frac{3}{4}$ as long. So this would be about $\frac{78}{100}$ as long. That's 78%!

**Ms. Rodriguez:** So the shorter side is about 78% as long as the longer one. Does that seem reasonable when you look at the picture? *(The students agree that it does.)* On your calculator enter "tan 38" and tell me what you notice.

**Ellen:** It equals about 0.781. That's really close to our average, 0.777.

**Ms. Rodriguez:** The number you calculated is called the "tangent" of 38°. It is the ratio of those two sides in a right triangle when the angle is 38°, so it's the answer we should have gotten. We were very close. Why do you think our answer is a little off?

**Jake:** I think it's because our drawings aren't exactly right.

**Ellen:** And our measurements aren't perfect, either.

**Ms. Rodriguez:** We did pretty well, though! You will learn more about what a *tangent* is when you study trigonometry. For now, the important thing to understand is that the ratio is the same for all of the triangles and that there's a way to predict it on your calculator!

*(The class continues by exploring the triangle with the 61° angle in the same way, but now it goes a little faster, because they know what to expect.)*

# Problem #4

Now that you know more about how the sides in your right triangles are related, you can investigate the height of the tower without using your drawing!

## *Directions*

- Use your value for tan38° to write an algebraic equation for one of your right triangles.
- Use tan61° to write an algebraic equation for the other right triangle.
- Use your equations to search for a more precise answer for the height of the tower.
- Compare your answer to your prediction in Problem #2. How did you do?
- Decide whether or not your answer is now exact. Explain your thinking.

## *Diving Deeper*

Is the relationship between an angle of elevation and the height of the tower proportional?

# CONVERSATION STARTERS FOR #4

*What do you notice? What do you wonder?*

*I wonder* how many variables I should use in my equations?

*I notice* that there are three side lengths ($H$, $x$, and $y$) involved in the two tangent comparisons.

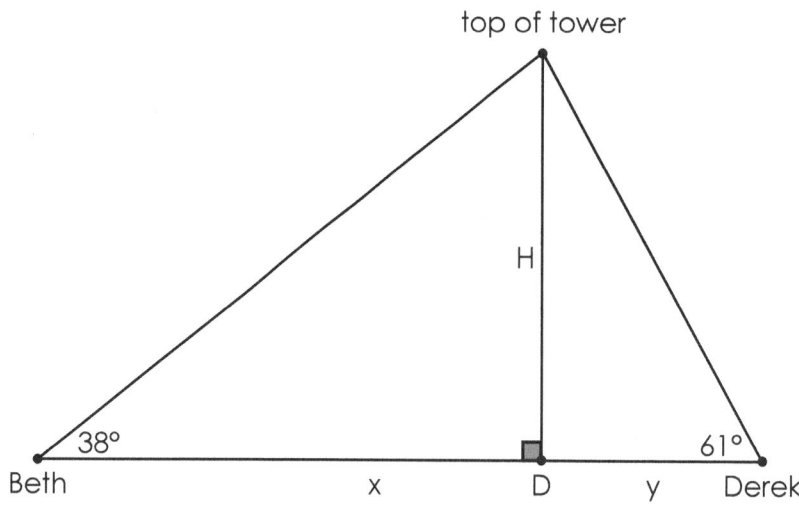

*I notice* that $x$ and $y$ have a simple relationship: $x + y = 88$ (or $y = 88 - x$).

*I wonder* how I can find the value of $H$ when there are so many values I don't know?

If you choose a value for one of the variables, can you figure out what the others have to be?

# SOLUTIONS FOR #4

*Equations for the right triangles*:

$$\frac{H}{x} \approx 0.78129 \qquad \frac{H}{y} \approx 1.80405$$

*One strategy for finding H*: Choose a value for $H$. Test and revise it as needed. If $H = 48.2$, then $x \approx 48.2 \div 0.78129 \approx 61.69$ and $y \approx 48.2 \div 1.80405 \approx 26.72$. This looks promising, because $x + y \approx 61.69 + 26.72 = 88.41$ meters, which is close to 88 meters. Because the total is slightly too large, we could decrease $H$ to make the total smaller.

$$
\begin{array}{ll}
H = 48.1 & x + y \approx 88.23 \\
H = 48.0 & x + y \approx 88.04 \\
H = 47.9 & x + y \approx 87.86
\end{array}
$$

*Comparison to problem #2*: Based purely on the calculations, the best answer looks like $H = 48.0$ meters. This fits well with the prediction from Problem #2.

*Exactness of the answer*: The calculations showed that our pictures were quite accurate! However, they probably did not improve the accuracy of our answer. If Beth and Derek's measurements were off by a small amount, it could change the answer by at least a couple of meters. It is still probably best to report the height of the tower as about 48 to 50 meters. Beth was probably right, but it is still hard to be sure!

# ALGEBRA CONNECTIONS

Algebra students may challenge themselves to solve the system of equations in Problem #4 using the method of substitution. Here is one approach.

$$\frac{H}{x} \approx 0.78129 \qquad \frac{H}{y} \approx 1.80405 \qquad x + y = 88$$

Solve the first two equations for $x$ and $y$:

$$x \approx \frac{H}{0.78129} \qquad y \approx \frac{H}{1.80405}$$

Substitute these expressions into the third equation:

$$\frac{H}{0.78129} + \frac{H}{1.80405} \approx 88$$

Multiply both sides by $(0.78129)(1.80405)$:

$$1.80405H + 0.78129H \approx 88(1.80405)(0.78129)$$

Add like terms and calculate the right side:

$$2.58534H \approx 124.03479$$

Divide both by sides by 2.58534:

$$H \approx 48.0 \text{ m}$$

Notice that we kept plenty of decimal places in the calculation process and rounded appropriately at the end.

# Exploration **6**

## Keep It in Proportion

In this exploration, students study proportional relationships in more depth by investigating *cross products*. In Stage 1, they create multiple strategies to make two fractions equivalent when one or two of the numbers are unknown. From this, they eventually discover that the cross products of a proportion are always equal, and they learn why.

In Stages 2 and 3, students explore relationships that appear to be proportional but are not. This prepares them to recognize these types of situations when they solve problems. Students also learn that every proportion that they solve is part of a larger proportional relationship. And conversely, if the original relationship is not proportional, you cannot write a correct proportion from it.

It may seem strange that I have waited until Exploration 6 to introduce cross products. I have found that when I wait, students learn to use what they know about equivalent fractions to explore all of the relationships between the numbers in a proportion. This makes them more flexible thinkers, and it helps them understand where the *cross products* idea comes from and why it makes sense. It is exciting to hear the ideas that students come up with when I allow them to think more independently.

# STAGE 1

Some of your students may already have learned about *cross products* from parents, teachers, or others. If so, ask them to be creative and to use other ways of thinking about these problems—just for now. Students will discover and explain the "what" and the "why" behind cross products at the end of Stage 1!

*A quick tip*: The Solutions to Problem #1 include a lot of strategies. Don't worry about reading through all of these right away. Just know that they are there for your reference. These are *students'* strategies! As your students share their ideas, keep your eyes open for connections. For example, comparing Strategies 5 and 6 in the Solutions shows that multiplying by $\frac{3}{11}$ has the same effect as dividing by $3\frac{2}{3}$, reinforcing what students already know about multiplying by the reciprocal of the divisor!

Do not be discouraged if your students don't produce a lot of ideas right away. Give them plenty of time, and realize that it may take them a while to get used to thinking independently and creatively about the calculations, especially if they are accustomed to being given the steps!

## What Students Should Know

> » Understand the meaning of *equivalent fractions*.
> » Multiply and divide fractions and decimals.

## What Students Will Learn

> » Use their understanding of equivalent fractions to solve challenging proportions.
> » Discover that cross products of proportions are equal and explain why.

# Problem #1

(a) $\dfrac{4}{7} = \dfrac{10}{y}$    (b) $\dfrac{x}{20} = \dfrac{3}{11}$

## Directions

- Solve the equations. Show and explain your thinking.
- Solve again using a different strategy! Show and explain your thinking.

## Testing the Waters

Answer the questions using the equations

(a) $\dfrac{20}{30} = \dfrac{x}{6}$ and (b) $\dfrac{9}{y} = \dfrac{6}{10}$.

# CONVERSATION STARTERS FOR #1

*What do you notice? What do you wonder?*

*I notice* that $y$ is not a whole number.

*I wonder* if it makes sense to have a fraction or decimal in the numerator or denominator?

*I notice* that $y$ must be greater than 10.

*I wonder* if it might help to write $\dfrac{4}{7}$ as an equivalent fraction?

*I notice* that 10 is 2.5 times greater than 4.

*I notice* that 7 is 1.75 times greater than 4.

*I notice* that 4 is $\dfrac{4}{7}$ as large as 7.

Similar questions and observations may apply to the second equation. For example: *I wonder* what to multiply 11 by in order to get 20?

# SOLUTIONS FOR #1

*Answers*: $y = 17.5$ and $x = 5\dfrac{5}{11}$ or $5.\overline{45}$. Sample strategies are shown below. All strategies will work for either equation.

*Strategy 1 (for a)*: Notice that $4 \cdot 2.5 = 10$. Multiply 7 by 2.5 to get 17.5.

*Strategy 2 (for a)*: Write $\dfrac{4}{7}$ as $\dfrac{20}{35}$. Because 10 is half of 20, take half of 35 to get 17.5.

*Strategy 3 (for a)*: Because 7 is $1\dfrac{3}{4}$ times as large as 4, multiply 10 by $1\dfrac{3}{4}$ to get $17\dfrac{1}{2}$.

*Strategy 4 (for b)*: $(20 \div 11) \cdot 3 = 5.\overline{45}$. ( $20 \div 11$ tells you what to multiply 11 by to get 20, so you multiply 3 by the same thing.)

*Strategy 5 (for b)*: 3 is $\dfrac{3}{11}$ of 11, so find $\dfrac{3}{11}$ of 20: $20 \cdot \dfrac{3}{11} = \dfrac{60}{11} = 5\dfrac{5}{11}$.

*Strategy 6 (for b)*: $20 \div 3\dfrac{2}{3} = 20 \div \dfrac{11}{3} = 20 \cdot \dfrac{3}{11} = \dfrac{60}{11} = 5\dfrac{5}{11}$. This works because 11 is $3\dfrac{2}{3}$ times greater than 3, so you need to know what to multiply by $3\dfrac{2}{3}$ to get 20.

# Problem #2

$$\frac{9}{b} = \frac{c}{6}$$

## Directions

- Find at least six solutions to the equation.
- Describe the relationship between $b$ and $c$.
- Change the 9 and the 6 to different numbers. Test that your relationship still holds.
- Finish this equation to describe an important property of proportions:

$$\text{If } \frac{a}{b} = \frac{c}{d} \text{ then } \underline{\hspace{2cm}}.$$

- Explain why the equation is always true (when $b \neq 0$ and $d \neq 0$).

# CONVERSATION STARTERS FOR #2

*What do you notice? What do you wonder?*

*I notice* that there are two simple solutions that use the numbers 6 and 9.

*I notice* that it helps to choose a value for one variable first and then find the other.

*I notice* that neither $b$ nor $c$ can equal 0.

*I notice* that it is easier to begin with factors of 9 or 6.

*I notice* that it helps to pay attention to how I change the 9 when I fill in $b$.

*I notice* that as $b$ gets larger, $c$ gets smaller.

*I notice* that if I make $b$ greater than 9, then I must make $c$ less than 6.

*I wonder* what happens to $c$ if I make $b$ twice as large as 9?

*I notice* that if I reverse the order of $b$ and $c$, I always get another solution!

*I wonder* how many solutions there are?


## After Students Have Discovered that
## b and c Have a Product of 54:

*I wonder* if *every* pair of numbers with a product of 54 is a solution to the equation?

# SOLUTIONS FOR #2

*A few solutions (written as pairs b, c):*

$$9, 6 \quad 3, 18 \quad 1, 54 \quad 18, 3 \quad 6, 9 \quad 4.5, 12 \quad 12, 4.5$$

*A relationship between b and c:* $b \cdot c = 54$, which is the same as $9 \cdot 6$!

*Changing the numbers:* The relationship still holds. For example, if $\dfrac{10}{b} = \dfrac{c}{8}$, some solutions are:

$$10, 8 \quad 5, 16 \quad 2, 40 \quad 1, 80 \quad 2.5, 32 \quad 16, 5 \quad 8, 10$$

$b \cdot c = 80$ for all pairs. The original numbers, 10 and 8 have the same product.

*A property of proportions:* If $\dfrac{a}{b} = \dfrac{c}{d}$, then $a \cdot d = b \cdot c$. A vocabulary note: $a \cdot d$ and $b \cdot c$ are called *cross products.*

*Why the property is true:* Think of $a$ and $d$ as the numbers you know. If you make $b$ twice as large as $a$, then you make $c$ half as large as $d$ (to keep the fractions equivalent). The two changes compensate for each other! You can prove this with variables:

$$b \cdot c = (2 \cdot a) \cdot \left( \frac{1}{2} \cdot d \right) = \left( 2 \cdot \frac{1}{2} \right) \cdot (a \cdot d) = 1 \cdot a \cdot d = a \cdot d$$

The compensation idea still works when you make other changes to $a$ and $d$. Try it! Other reasons are possible. Some students may see that equivalent fractions can always be written as $\dfrac{a}{b}$ and $\dfrac{a \cdot n}{b \cdot n}$, and that these have the same cross products.

# STAGE 2

Most students benefit from extra practice using cross products before they begin Stage 2. You can assign some exercises from their math text, or encourage them to go back to some of the problems from Explorations 1–4 and solve them with cross products.

Before you start Problem #3, introduce the word *proportion* (two equivalent ratios or equal rates), and review the term *cross products* from Problem #2. Remind students that in a *proportional relationship*, (1) the pairs form equivalent ratios, and (2) there is a constant rate of change, and both quantities begin together at 0.

Problem #4 is a continuation of Problems #6 and #7 in Ramps, Paints, and Hot Air Balloons, but students can work on it whether they have solved the earlier problems or not.

## What Students Should Know

» A *proportional relationship* is a set of equivalent ratios or equal rates.

## What Students Will Learn

» Use *cross products* to solve proportions.
» Reinforce understanding that a *proportional relationship* has a constant rate of change and begins with both quantities equal to 0.
» Understand that every proportion is part of a larger proportional relationship.
» Recognize relationships that are not proportional, and realize that you cannot write proportions from them.

# Problem #3

Justin lives of the 8th floor of an 11-floor apartment building. He has just learned about proportions, and he is excited to try out his new knowledge. He writes and solves the equation $\frac{8}{112} = \frac{11}{n}$, and then he counts the steps to the top of his building. He is surprised!

## Directions

- Explain why Justin was surprised.
- Explain how to resolve Justin's dilemma.
- Create two different tables that show the situation correctly.
- Compare the tables, and expand on your explanation.

# CONVERSATION STARTERS FOR #3

*What do you notice? What do you wonder?*

*I wonder* what the number 112 means in Justin's equation?
> He climbs 112 stairs to reach the 8th floor.

*I wonder* if all flights have the same number of stairs?
> Is Justin assuming that they do? How would it affect the problem if they did not? (Moving forward, you may assume that they do.)

*I wonder* how many steps Justin counted?
> Try counting them yourself! (We will suppose that he counted them correctly!)

*I wonder* why Justin didn't just count the number of stairs per flight?
> He probably did, but he was still curious why his equation didn't work!

*I notice* that it helps to make a table(s) right away.
> Yes—tables help you organize and clarify your thinking!

*I notice* that Justin doesn't have to climb any stairs to get to the first floor!

*I wonder* if I can change Justin's equation to make it work?

*I wonder* how it is possible to create two different correct tables for one situation?
> Try changing one of the variables.

*I wonder* what variables to use?
> What caused Justin's dilemma in the first place?

*I notice* that pairs of numbers in one of the tables do not make equivalent fractions.

*Note*: If students are not sure how to use the tables to expand upon their explanation, suggest that they talk about *proportional relationships*.

# SOLUTIONS FOR #3

Justin was trying to calculate the number of steps to reach the top floor. He was surprised because the solution to his equation (154) was different than the number of steps he counted (160).

*Resolving Justin's dilemma*: The number of floors is not the same as the number of flights of stairs, because the bottom floor is the 1st floor, not the "0th" floor. Justin climbs 7 flights of stairs to reach the 8th floor and 10 flights to reach the 11th floor. His equation should be:

$$\frac{7}{112} = \frac{10}{n}$$

Solving with cross products: $7 \cdot n = 10 \cdot 112$, $7 \cdot n = 1120$, $1120 \div 7 = 160$

*Two correct tables*:

| Floors | 1 | 2 | 3 | 4 | 5 | 6 | 7 | 8 | 9 | 10 | 11 |
|--------|---|----|----|----|----|----|----|-----|-----|-----|-----|
| Stairs | 0 | 16 | 32 | 48 | 64 | 80 | 96 | 112 | 128 | 144 | 160 |

| Flights | 0 | 1 | 2 | 3 | 4 | 5 | 6 | 7 | 8 | 9 | 10 |
|---------|---|----|----|----|----|----|----|-----|-----|-----|-----|
| Stairs | 0 | 16 | 32 | 48 | 64 | 80 | 96 | 112 | 128 | 144 | 160 |

*Comparing the tables*: The table of floors and stairs is correct, but it is not a proportional relationship. That is, the pairs do not form equivalent ratios. For example, 2 : 16 is not equivalent to 3 : 32. The problem is that this table does not begin with both values equal to 0.

When Justin tries to write a proportion for the floors and stairs, it changes the number of stairs per flight. The table fits his calculation, but not reality!

| Floors | 1 | 2 | 3 | 4 | 5 | 6 | 7 | 8 | 9 | 10 | 11 |
|--------|----|----|----|----|----|----|----|-----|-----|-----|-----|
| Stairs | 14 | 28 | 42 | 56 | 70 | 84 | 98 | 112 | 126 | 140 | 154 |

# Problem #4

| | Temperature and Volume of Air (Constant Pressure) | | | | | |
|---|---|---|---|---|---|---|
| Temperature (°K) | 266 | 273 | 301 | 320 | 345 | 355 |
| Volume (mL) | 4.5 | 4.6 | 5.1 | 5.4 | 5.8 | 6.0 |

| | Temperature and Volume of Air (Constant Pressure) | | | | | |
|---|---|---|---|---|---|---|
| Temperature (°C) | –7 | 0 | 28 | 47 | 72 | 82 |
| Volume (mL) | 4.5 | 4.6 | 5.1 | 5.4 | 5.8 | 6.0 |

Kiab's equation: $\dfrac{5.1}{301} = \dfrac{V}{400}$    Azar's equation: $\dfrac{5.1}{28} = \dfrac{V}{127}$

## Directions

- Describe the quantity that the girls are trying to calculate.
- Find the solution to each equation.
- Tell which girl's answer is correct, and explain the error that the other made.

# CONVERSATION STARTERS FOR #4

*What do you notice? What do you wonder?*

*I wonder* what the data in the tables are about?
> A small sample of air was heated in order to observe its expansion. (See Exploration 2.)

*I wonder* if the two tables show the same measurements?
> Yes, the temperature measurements are simply expressed in different units.

*I wonder* what °K and °C mean?
> They are Kelvin and Celsius units of temperature. How do they compare?

*I wonder* why anyone would use a temperature scale with such large numbers?

*I notice* that the fractions on the left sides of the equations come directly from the tables.

*I wonder* why the denominators on the right sides of the equations are different?
> How do 400°K and 127°C compare?

*I notice* that the $V : T$ ratios in the first table have nearly the same rate.

*I notice* that the first $V : T$ ratio in the second table is negative!

*I wonder* what temperature results in a volume of 0 mL?

*I wonder* if graphs would help me understand these relationships better?

*I notice* that Kiab's solution seems much more reasonable than Azar's.

# SOLUTIONS FOR #4

*The quantity*: The girls are calculating the volume of air when the temperature is 400°K (127°C).

*The solutions to each girl's equation*:

» Kiab's equation: $\dfrac{5.1}{301} \text{ mL} = \dfrac{V}{400}$, $301 \cdot V = 5.1 \cdot 400 = 2040$, $2040 \div 301 \approx 6.8$

» Azar's equation: $\dfrac{5.1}{28} \text{ mL} = \dfrac{V}{127}$, $28 \cdot V = 5.1 \cdot 127 = 647.7$, $647.7 \div 28 \approx 23.1$

Kiab's solution obviously appears to fit the data better.

*The correct solution*: Kiab's solution is the correct one. Azar's table does not represent a proportional relationship, so it does not make sense to use it to write a proportion.

There are two ways to see why the relationship in Kiab's table is proportional, but Azar's is not. First, the pairs in Kiab's table form fractions that are (approximately) equivalent, because they all have about the same rate (0.0169 mL per 1°K). The fractions from the pairs in Azar's table do not.

Second, in Kiab's table, 0 mL matches with 0°K. That is, both values begin at 0. (See the Solutions to Problem #7 in Ramps, Paints, and Hot Air Balloons). For Azar's table, 0 mL corresponds to −273°K. Even though both tables correctly show the same physical data, one relationship is proportional, and the other is not. This explains why scientists often use the Kelvin scale!

# STAGE 3

In Problem #5, students explore the math of objects in balance. They discover that even though they can represent the situation with two equal fractions, it is not a proportional relationship, because each input/output pair has the same product rather than the same quotient.

Although students can often predict how to balance the objects, they enjoy making mobiles to test their ideas! A hanger with a bag of marbles tied to each end works well.

### What You Will Need

» Hangers, string, food storage bags, marbles, or dice (all optional).

### What Students Should Know

» Represent real-world data with tables, graphs, and formulas.
» Understand concepts from Stage 1 of this exploration.

### What Students Will Learn

» Reinforce understanding of proportional relationships.
» Begin to explore *inverse proportionality*.

# Problem #5

Leonard makes a *mobile* by hanging two bags, each containing 8 marbles, at distances of 12 centimeters from the center of a light rod. (Each marble has the same weight.)

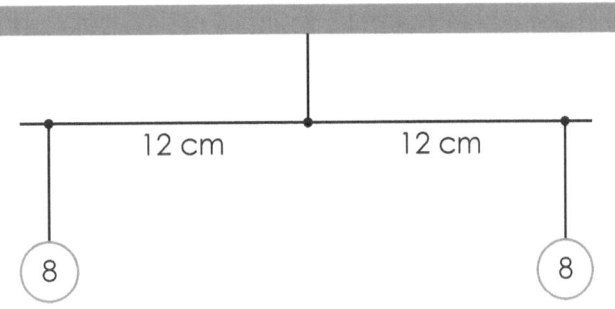

12 cm        12 cm

8          8

## Directions

- Leaving one side as it is, change the weight and distance on the other side so that the mobile remains balanced. Make a table of weight-distance combinations that work. Explain your thinking strategies.
- Describe any patterns that you can find in your table.
- Graph the weight-distance relationship, and find a formula for it.
- Determine whether the relationship is proportional. Explain your thinking.
- Decide if it is possible to write an equation of equivalent fractions for this situation. If so, do it. If not, explain why it cannot be done.

## Diving Deeper

Think of other real-world situations in which the same mathematical patterns occur.

# CONVERSATION STARTERS FOR #5

*What do you notice? What do you wonder?*

*I notice* that I don't have to know how heavy each marble is.

> In fact, because the marbles are all the same weight, you can use *marbles* as your unit of weight!

*I notice* that if I add marbles to a bag, I have to move the bag closer to the center.

> In other words, when the weight increases, the distance decreases.

*I wonder* what happens if I double the weight on one side?

*I notice* that balancing the mobile involves reciprocals!

*I notice* that the weight becomes very large as the bag gets very close to the center.

*I notice* that the distance becomes very large when the weight gets close to 0.

*I notice* that instead of equal rates and cross products, my table shows equal products and "cross rates"!

*I wonder* what the graph looks like if you extend it to the left and the right?

# SOLUTIONS FOR #5

*Some weight-distance combinations*:

| Marbles (m) | 1 | 2 | 4 | 8 | 12 | 16 | 20 | 24 |
|---|---|---|---|---|---|---|---|---|
| Distance (d) | 96 cm | 48 cm | 24 cm | 12 cm | 8 cm | 6 cm | 4.8 cm | 4 cm |

Multiply 8 marbles by some number. Then divide 12 cm by the same number (or multiply by its reciprocal). For example, if you double the 8 marbles, then you must halve the distance to the center of the rod. Or, if you multiply the number of marbles by 1.5, then you multiply the distance from the center by $\frac{2}{3}$.

*Patterns in the table*: The number of marbles times the distance is always equal to the same number, 96. (Also, instead of equal cross products, there are equal "cross rates!")

*Graph*:

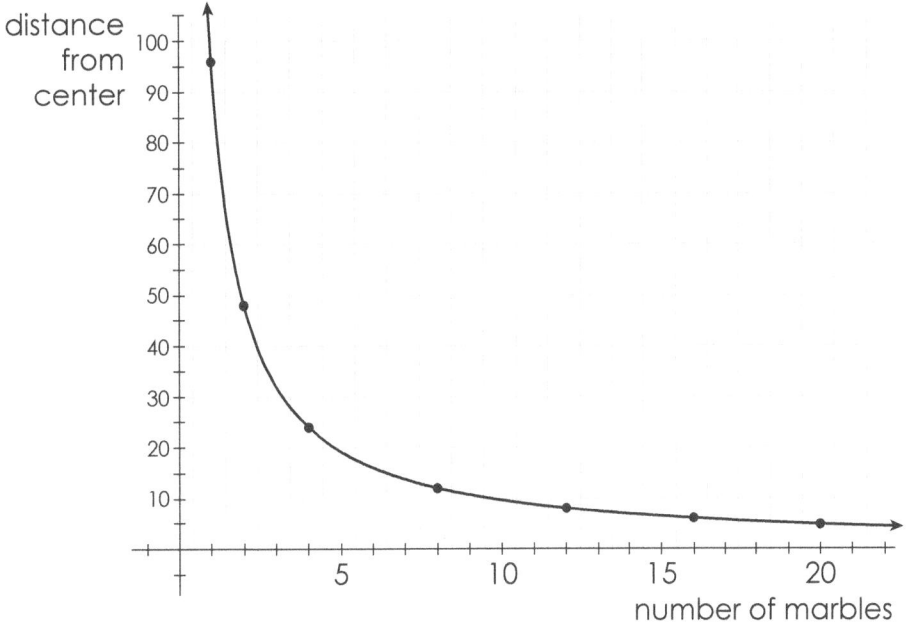

Students may or may not connect the points with a curve. Connecting them shows what would happen with "fractional marbles."

*Possible formulas*:

$$d = \frac{96}{m}, \ m = \frac{96}{d}, \ d \cdot m = 96$$

*Description of the relationship:* The relationship is not proportional, because the pairs in the table do not form equal rates. (However, they do have equal products!) Also, the graph of a proportional relationship is a straight line beginning at the *origin* [the point $(0, 0)$]. This graph obviously does not fit that pattern!

*Writing the relationship with equivalent fractions:* Surprisingly, it is possible to write the relationship with equivalent fractions! For example, you may write $\dfrac{1}{48} = \dfrac{2}{96}$ or $\dfrac{1}{2} = \dfrac{48}{96}$. However, you may not write $\dfrac{1}{96} = \dfrac{2}{48}$.

This type of relationship is known as *inversely* proportional. Some key features of inversely proportional relationships are:

» When one quantity is multiplied by a number, $n$, the other is divided by $n$.
» The product of each input and output is a constant number.
» The graph is a *hyperbola* (the shape shown above).
» The equation has the form $y = \dfrac{k}{x}$ or $x \cdot y = k$, where $k$ is the constant product.

# ALGEBRA CONNECTIONS

In this exploration, students study cross products by applying their understanding of relationships between the numerators and denominators of equivalent fractions. In algebra courses, the standard way to prove that cross products of equivalent fractions are equal is to multiply both sides of the equation $\frac{a}{b} = \frac{c}{d}$ by the common denominator, $bd$ :

$$\frac{a}{b} \cdot bd = \frac{c}{d} \cdot bd$$

$$\frac{a}{b} \cdot \frac{bd}{1} = \frac{c}{d} \cdot \frac{bd}{1}$$

$$\frac{abd}{b} = \frac{cbd}{d}$$

The common factors of $b$ on the left and $d$ on the right divide out, and you are left with $ad = cb$. Cross products are especially helpful for solving proportions that involve algebraic expressions.

In Problem #5, prealgebra students expand their awareness of proportional relationships to include *inverse proportionality*. This shows the need for caution when writing the equations that relate two fractions. In algebra courses, students will learn to distinguish various types of direct and inverse proportional relationships. They will study formulas, tables, and graphs for some of them, and develop procedures for manipulating and solving the related algebraic equations.

# Exploration **7**

## Grab Bag

Each stage of this "grab bag" of problems extends your students' understanding of rates and proportions in some new way. In the first problem, students literally grab blocks from a bag one handful at a time and estimate the total number of blocks. Their challenge is to combine a large amount of data in a situation involving proportions. In Stage 2, students explore reciprocal rates, and they look at what happens when addition and subtraction are thrown into the mix. Finally, in Stage 3, they investigate average rates.

Each stage has a real-world connection, and you can adapt the problems by asking students to perform their own experiments or having them collect and analyze data from their own experience. I have given specific suggestions for doing this in Stages 1 and 2.

# STAGE 1

In Stage 1, your students estimate the number of blocks in a paper bag. The catch is that they are allowed to see only a handful of blocks at a time!

I often begin with the final question in the Conversation Starters. As students analyze the data, I ask them to pretend that they are counting fish in a lake. I have them think about how the process of gathering the data would look in the real world. How would they design it to obtain the best possible estimate? What might go wrong?

To make the problem more "hands-on," give your students a bag of blocks in which a known number of them are marked. Have them do the experiment and collect the data themselves. This will enable them to check their estimates by counting the blocks! Of course, you will have to work up your own answers, but you can still use the Solutions section to see some good strategies.

## What Students Should Know

  » Solve problems involving ratios and percentages.

## What Students Will Learn

  » Create and practice strategies for finding an unknown whole.
  » Compare methods for combining estimates to make a "best" prediction.
  » Explore the effects on calculations of small changes in data.
  » Make decisions about how to report estimates.

# Problem #1

A bag contains a collection of identical blocks. Without looking, you draw a handful of 24 blocks, place a mark on each one, return the blocks to the bag, and mix them thoroughly. Ten people in your class then take turns drawing a random set of blocks and counting the blocks and marked blocks in the set. The person who drew the blocks returns them to the bag and mixes them thoroughly before the next person draws.

| Person Number | 1 | 2 | 3 | 4 | 5 | 6 | 7 | 8 | 9 | 10 |
|---|---|---|---|---|---|---|---|---|---|---|
| Marked Blocks Drawn | 2 | 1 | 3 | 1 | 2 | 2 | 1 | 4 | 1 | 3 |
| Total Blocks Drawn | 13 | 22 | 30 | 15 | 17 | 30 | 18 | 20 | 26 | 34 |

## Directions

- Estimate the total number of blocks in the bag. Explain your thinking process, including calculations.
- Compare your answer to others in your class. Explain what causes any differences.

# CONVERSATION STARTERS FOR #1

*What do you notice? What do you wonder?*

*I notice* that there are many different strategies for estimating the number of blocks.

*I notice* that the estimates vary a lot!

*I wonder* why the estimates are so different?

*I wonder* if I can predict which estimate is more likely to be correct?

*I wonder* if I could use the average *percentage* to combine the estimates?

*I wonder* how it would affect the estimates if one person had drawn just one more marked block than they did?

*I wonder* if it matters which person drew the extra block?

*I wonder* how I could make an estimate if someone drew 0 blocks?

*I wonder* if I should report my estimate as a number or as an interval?

*I wonder* if I can apply what I learned to count something in real life?

Suppose you want to count the fish in a lake. Imagine catching some of the fish, marking them, and coming back later. How is this like the process of drawing blocks from the bag? How would the process work in the real world? What factors might affect the accuracy of your estimates? (Some students may be interested in researching the *capture-tag-recapture* technique for estimating animal populations.)

# SOLUTIONS FOR #1

Estimates should probably be in the upper 200s to the lower 300s. Some students will combine the data for all 10 people before calculating. Others will calculate a prediction for each person and then combine the results.

*Strategy 1* (combining the data first): Compare the total number of marked blocks to the total number of blocks drawn.

» Total number of marked blocks drawn: 20
» Total number of blocks drawn: 225

There are many methods for using these numbers to estimate the total number of blocks, $N$.

1. The ratios of marked blocks to total blocks should be approximately equivalent for the draws and for the bag:

   marked (draws) : total (draws) = marked (bag) : total (bag)

   $$20 : 225 = 24 : N$$

   Because you multiply 20 by 1.2 to get 24, you also multiply 225 by 1.2 to get 270.

2. The percentage of marked blocks should be approximately equivalent for the draw and for the bag. Because 20 is about 83.3% of 24, 225 it should be about 83.3% of $N$:

   $$225 \div 83.3 \approx 2.701 \text{ is approximately 1\% of } N$$

   $N$ is approximately $2.70 \cdot 100 = 270$ blocks

3. The total number of blocks drawn is approximately $225 \div 20 = 11.25$ times the number of marked blocks drawn. Therefore, $N \approx 24 \cdot 11.25 = 270$ blocks.

*Strategy 2* (calculate individual predictions first):
*Predictions for each person (using methods such as the three above):*

| | | | | |
|---|---|---|---|---|
| 1 | 156 blocks | | 6 | 360 blocks |
| 2 | 528 blocks | | 7 | 432 blocks |
| 3 | 240 blocks | | 8 | 120 blocks |
| 4 | 360 blocks | | 9 | 624 blocks |
| 5 | 204 blocks | | 10 | 272 blocks |

Students who follow this approach will have to decide on a way to combine the results for each person. Most choose to use the mean. Adding the numbers and dividing by 10 gives an estimate of about 330 blocks.

*I wonder* why Strategy 1 for combining estimates gives a smaller answer than Strategy 2?

*I wonder* which method is likely to be more accurate?

Think: In Strategy 2, each drawing is given equal weight, regardless of the number of blocks drawn.

*I wonder* if it would help to combine experiments one at a time (keeping a running total)?

It is interesting to see if your predictions gradually stabilize.

# STAGE 2

In Stage 2, students learn to use rates to measure how crowded a space feels. This involves the concept of *population density*, the number of people per unit area. In this activity, students may choose to use the reciprocal of the population density, the *area per person*, because it is easier to visualize.

In real-world scenarios like the one in Stage 2, we often simplify the situation in order to make it practical to do the mathematics. Encourage your students to be skeptical! They may believe that the area per person at a school is less important than how the space is laid out and used. They may also notice that some of the mathematical calculations seem too precise. For example, although moving 107 students from one school to another will equalize the area per person, it is not likely important for it to be exactly the same at each school.

Both of these observations suggest that we should take our results with a grain of salt. However, this does not mean that the calculations are useless! Conversations like these help students to learn about the strengths and limitations of mathematical models and the importance of making sense of your results once you get them.

## What Students Should Know

- » Understand and calculate rates.
- » Understand and solve proportions.

## What Students Will Learn

- » Understand and apply reciprocal rates.
- » Solve proportions whose quantities change by addition or subtraction.
- » Use calculations to make decisions.

# Problem #2

| Bristlecone School District Data | |
|---|---|
| Mountain Heights Middle School | 51,600 square feet of floor space |
| | 470 students |
| North Star Middle School | 118,300 square feet of floor space |
| | 725 students |

## Directions

- Explain why this data may have been gathered.
- Do a quick calculation for each school to illustrate the issue.
- Use math to recommend a plan to solve the problem.

# CONVERSATION STARTERS FOR #2

*What do you wonder? What do you notice?*

*I wonder* how it would feel to be at the Mountain Heights school?

*I wonder* how it would feel different at North Star Middle School?

After students have done a quick calculation for each school, introduce the term *population density* (number of people per unit area).

*I notice* that when the population density is greater, the school is more crowded.

*I notice* that when the area per person is greater, the school is less crowded.

*I notice* that the area per person is the reciprocal of the population density.

*I wonder* why people usually use population density instead of area per person?

## After Students Have Identified the Problem

*I wonder* how to make Mountain Heights Middle School less crowded?

*I wonder* what the area per person is at our school?

*I wonder* how the area per person compares at different schools in our district?

*I wonder* what the area per person is for a typical middle school?

*I wonder* what the area per person is in our classroom?

*Note*: *Area per person* could be replaced by *population density* in these questions.

# SOLUTIONS FOR #2

*Why the data may have been collected*: The school district is probably concerned that Mountain Heights Middle School is much more crowded than North Star Middle School.

*Some useful numbers*:

|  | *area ÷ population* | *population ÷ area* |
|---|---|---|
| **Mountain Heights Middle School** | 110 ft² per person | 0.0091 people per ft² |
| **North Star Middle School** | 163 ft² per person | 0.0061 people per ft² |

*Note*: The number of people per unit area is called the *population density*, which is the usual measure for "crowdedness." However, the area per person may be easier for students to visualize in this situation.

*One way to solve the problem (see Problem #3 for a different approach)*: Expand Mountain Heights Middle School. The school would need to add a little more than 25,000 square feet in order to equalize the number of square feet per person. Some students may write a proportion to find this number:

$$\frac{A}{470} = \frac{118,300}{725}$$

They may use cross products to obtain 76,700 square feet. Subtracting the original area gives the result:

$$76,700 - 51,600 = 25,100 \text{ square feet}$$

Another approach is to begin by multiplying 163 (the number of square feet per person at North Star) by 470 (the population of Mountain Heights). This gives a slightly different answer because the number 163 was rounded.

# Problem #3

| Bristlecone School District Data | |
|---|---|
| Mountain Heights Middle School | 51,600 square feet of floor space |
| | 470 students |
| North Star Middle School | 118,300 square feet of floor space |
| | 725 students |

## Directions

- Use math to describe a second method to solve the overcrowding problem.
- Choose between the two options, or identify other information you would need before deciding. Explain your reasoning based on the data and your calculations.

# CONVERSATION STARTERS FOR #3

*What do you notice? What do you wonder?*

*I wonder* how much it would cost to expand the Mountain Heights building?

*I wonder* if the district has enough money for the expansion?

*I wonder* what the district can do if they don't have money for building projects?

*I notice* you could improve the crowding by moving students.

*I wonder* how many students would have to move to solve the problem?

*I wonder* if I can find the number of students without using trial and error?

*I wonder* how it affects the area per person each time one person moves?

*I wonder* if it would help to imagine putting the two schools together?

*I wonder* if there are other things that might make a school building feel crowded?

*I wonder* how much the crowding affects students' learning?

# SOLUTIONS FOR #3

*Another solution to the Mountain Heights Middle School challenge*: Change the school attendance boundaries to move students from Mountain Heights to North Star. The number of students who must move in order to equalize the area per person is about 107.

Because the number of students leaving Mountain Heights is equal to the number of students entering North Star, a proportion for the situation looks like this:

$$\frac{51,600}{470-x} = \frac{118,300}{725+x}$$

$x$ represents the number of students who move. Cross products may not be very helpful here unless students know how to solve algebraic equations. Trial and error is a common strategy. Some students may have more creative ideas. For example:

- » $118,300 + 51,600 = 169,900$ square feet
- » $470 + 725 = 1195$ people
- » $169,900 \div 1195 \approx 142.176$ square feet per person
- » $51,600 \div 142.176 \approx 363$ people in Mountain Heights
- » $470 - 363 = 107$ students who move

Before reading ahead, see if you can picture the thinking behind these calculations! Students may have imagined joining the two school buildings and spreading their combined population out so that each person occupies the same amount of floor space. Then, when you "separate" the schools again, each one has the right number of students!

Ask students how they can verify their answers. They may check that the number of students leaving one school is equal to the number entering the other and that the number of people per square foot is now about the same at both schools.

*I wonder* why the number of square feet per person in this calculation is not equal to its average for the two schools?

*The better option*: Students may have different opinions. 107 students is about 23% of the 470 students at Mountain Heights! Asking nearly a quarter of the student body from Mountain Heights to change schools would be disruptive to the school and to many families. If the district can obtain the funds to expand the building, that may be the better approach. On the other hand, the number of classrooms in each building may be more important than the total floor space. Mountain Heights might be able to find ways to use its existing space more effectively.

# CLASSROOM VIGNETTE FOR #3

Mr. Hill's students have just solved Problem #3 using a guess and test strategy. He is ready to move on to another problem when Katelyn raises her hand with a question.

**Katelyn:** How do we know what the area per person will be after you make it the same at both schools?

*(Realizing that it may lead to a more efficient strategy to solve the problem, Mr. Hill changes his plan and decides to follow up on Katelyn's question.)*

**Mr. Hill:** Does anyone have any ideas?

**Jamal:** I don't understand Katelyn's question. Don't we already know what it is because we solved the problem?

**Katelyn:** Yeah, but could we have figured that part out before we got the answer?

**Mr. Hill:** I think that Katelyn's question might help us think about the problem in a new way! Can you think of a way to find the area per person that makes it the same at both schools without guessing and testing numbers?

*(Mr. Hill waits for a minute or so. There is no response, but there are a quite a few people who look like they have ideas that they are not quite prepared to share.)*

**Mr. Hill:** Take a couple of minutes to talk to your partner about the question.

*(As Mr. Hill listens, he notices that they do have some ideas. He hears two plans emerging.)*

**Mr. Hill:** I heard some interesting ideas. Who would like to share?

**Jasmine:** Well, I just put the two schools together in my head. So I added the two areas and the two populations. I got 169,900 square feet and 1195 people. Then I divided those and I got about 142.2.

*(Mr. Hill records these calculations on the board.)*

**Mr. Hill:** What does this number tell you? 142.2 what?

**Jasmine:** 142.2 square feet per person. *(Mr. Hill writes the unit.)*

**Mr. Hill:** Does anyone have a different strategy?

**Jamal:** We just averaged the two numbers. Mountain Heights was 110 square feet per person and North Star was 163, so we added them up and divided by 2. We got 136.5.

**Mr. Hill:** *(Records Jamal's process)* Does anyone have a name for what Jamal just did?

**Jasmine:** He did the mean of the two numbers!

**Mr. Hill:** *(Pointing to the numbers on the board)* So we have two answers! One of them is the mean, and the other is greater than the mean. Can they both be right?

**Katelyn:** I don't think they can both be right because there's a definite answer.

**Mr. Hill:** How can we decide if either of them is correct?

**Katelyn:** The mean makes more sense to me because it's in between.

**Jamal:** I like the mean, too, because we just got that, but I think Jasmine's answer is closer to what we got when we actually solved the problem.

**Mr. Hill:** Is that true? *(The class looks back at their earlier work and nods.)* So it looks like 142.2 might be right, but it seems strange to some people that it's not the mean. *(Again, most students nod.)*

**Mr. Hill:** Okay, let's pretend that we're starting fresh, and we haven't solved the problem yet. Think about the mean. How does it compare to the 110 and 163?

**Katelyn:** It's between them?

**Mr. Hill:** *(Waits for a minute and then asks)* Can we say more?

**Jasmine:** Is it *exactly* in the middle?

*(The class discusses the meaning of "mean" for a few minutes and agrees that Jasmine is correct. The other number, 142.2, is closer to 163 than to 110.)*

**Mr. Hill:** So for some reason, it looks like 142.2 is the right answer, even though it's closer to 163 than to 110? *(The class nods in agreement.)* How can we check this?

**Jamal:** Since we're starting over, maybe we can use 142.2 to get the answer to the number of kids that move.

**Mr. Hill:** How could we do that? *(The class thinks for a while, but they are struggling.)* Okay, please take some time with your partner to use the 142.2 square feet per person to figure out how many kids should move. *(After a few minutes observing, he notices an error that will be helpful to discuss, so he brings the class together.)*

**Mr. Hill:** I'm not sure if anyone is finished, but let's talk about what you have so far. Do you mind sharing, Jamal?

**Jamal:** Well, I'm not sure if this works, but I think we tried 51,600 times 142.2. *(He does it on his calculator as he is speaking.)* Wait—that's way too big! That would be like 7 million people at the school!

**Mr. Hill:** *(After waiting for a few moments for a response)* Can anyone help?

**Jasmine:** Maybe we should divide so it's not so big.

*(The class tries this and has a discussion about why it makes sense that the area divided by the area per person should be the number of people at Mountain Heights. They get an answer of about 363 people.)*

**Mr. Hill:** How can we check that 363 people is correct?

**Katelyn:** We could subtract it from the kids already there to find out how many moved and then do the same thing for North Star to see if the same number of kids move.

*(The students carry out Katelyn's idea and find that their solution is correct. They continue by talking about why the correct area per person should be closer to 163 than to 110.)*

# STAGE 3

The two problems in Stage 3 are about average rates. The calculations are not especially hard, but it is not so easy to figure out what "average speed" should mean in this situation. The Conversation Starters may help you lead a discussion about finding a sensible meaning instead of relying on a procedure.

A quick tip: If you ask students to formulate their own questions for Problem #4 before you give them the Directions on the Problem page, they may ask how long the trip will take. If so, include this as one of their tasks. Doing this may spur them to think of "average speed" in a more helpful way.

## What Students Should Know

- » Add, subtract, multiply, and divide fractions and decimals.
- » Understand and calculate rates.

## What Students Will Learn

- » Solve challenging problems with rates.
- » Think carefully when finding average rates.
- » Explore patterns of change in average rates.
- » Use graphs to represent average rates.

# Problem #4

Anna's family is driving 80 miles to her grandparents' house. They drive the first half of the trip at 30 miles per hour. The speed limit increases, and they drive 50 miles per hour for the rest of the trip.

## Directions

- Predict or estimate their average speed for the trip.
- Calculate their average speed for the trip.
- Find the average speed if the distance to her grandparents' house is 160 miles.
- Find a way to predict the average speed for any distance. Explain your thinking.

# CONVERSATION STARTERS FOR #4

*What do you notice? What do you wonder?*

*I wonder* what "average speed" means here? (Why would someone want to know it?)

*I wonder* if "average speed" means something different than "average of the speeds"?

*I notice* that Anna's family spends more time traveling at the lower speed.

*I wonder* what fraction of the time the family spends at each speed?

*I wonder* how long it takes the family to get to her grandparents' house?

*I notice* that if the distance changes, the fraction of the time spent at each speed does not change.

# SOLUTIONS FOR #4

The average speed for the trip is 37.5 miles per hour.

*Estimates or predictions*: Students' first guess is usually 40 miles per hour, because 40 is the mean of 30 and 50. Some may realize that the average time should be less than this, because the family spends more time driving 30 miles per hour than 50 miles per hour.

*The average speed (80 mile distance)*: The average speed is the constant speed the family would have to drive in order to arrive at the same time. In other words, it is the total distance divided by the total time.

» Total distance: 80 miles

» Time for the first half: 40 miles ÷ 30 miles per hour $=1\frac{1}{3}$ hours (80 min)

» Time for the second half: 40 miles ÷ 50 miles per hour $=\frac{4}{5}$ hours (48 min)

» Total time: $1\frac{1}{3}+\frac{4}{5}=2\frac{2}{15}$ hours (128 min)

» Average time: $80\div\frac{32}{15}=80\cdot\frac{15}{32}=\frac{75}{2}=37.5$ miles per hour

*The average speed (160 mile distance)*: Using the same process as above, students will find that the average speed is still 37.5 miles per hour (assuming half the trip at each speed).

*The average speed (any distance)*: The average speed will be 37.5 mph for any distance. For example, when the distance doubles, so does the total travel time, leaving the average speed unchanged. Some students may observe that for any distance, they spend the same fraction of the time at each speed.

# Problem #5

Anna's family is driving 80 miles to her grandparents' house. They drive the first half of the trip at 30 miles per hour. The speed limit increases, and they drive 50 miles per hour for the rest of the trip.

You can use graphs to gain new insights into real-world situations.

## Directions

- Draw a time versus speed graph to show what happened on Anna's trip.
- Draw a time versus distance graph to show what happened on Anna's trip.
- Explain in detail how each graph represents the situation.
- Show the average speed on at least one of your graphs. Explain your thinking.

# CONVERSATION STARTERS FOR #5

*What do you notice? What do you wonder?*

*I wonder* which variable should go on the horizontal axis?

Time is usually the independent variable, which goes on the horizontal axis.

*I notice* that it would also be possible to draw a distance versus speed graph.

## The Time Versus Speed Graph

*I notice* that the jump in the graph shows the speed changing instantly.

This isn't really possible, of course. We are ignoring the details at the time the speed changed.

*I notice* that the car spends $\frac{3}{8}$ of its time at the higher speed.

Interestingly, the dotted line is also $\frac{3}{8}$ of the way between 30 and 50!

## The Time Versus Distance Graph

*I notice* that each piece of the graph is straight because the speed is constant during those times.

# SOLUTIONS FOR #5

*A time versus speed graph:*

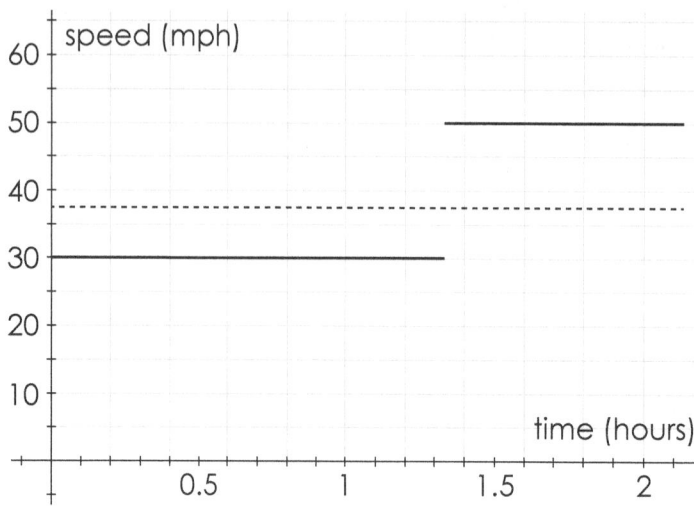

Each piece of the graph is flat, because the speed does not change. The dotted line shows the average speed. It is closer to 30 mph than 50 mph, because the car spent more time traveling at 30 mph.

*A time versus distance graph:*

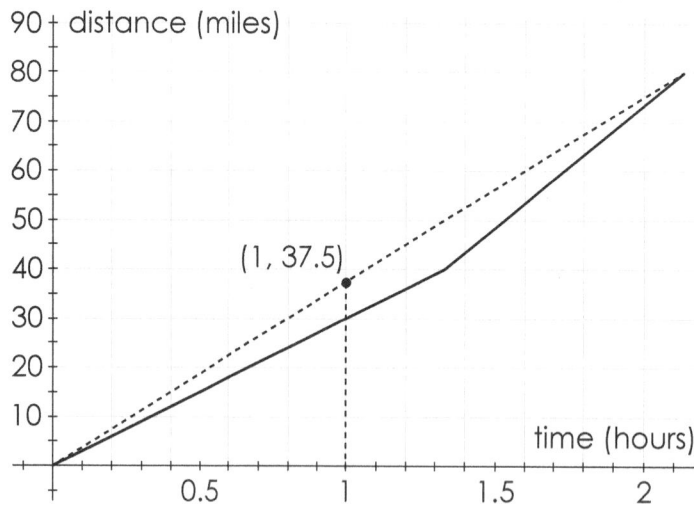

The steepness (slope) of each line represents the speed, which is the rate of change of the distance over time. The graph bends upward after $1\frac{1}{3}$ hours when the speed increases. The slope of the dotted line shows the average speed! This line shows that at a constant rate, the car would have traveled 37.5 miles after 1 hour.

*I wonder* how it affects each graph when you change the total distance?

## ALGEBRA CONNECTIONS

In Problems #2 and #3, algebra students may solve proportions using cross products and the distributive property. For example, Problem #3 might look like this:

$$\frac{51,600}{470-x} = \frac{118,300}{725+x}$$

$$51,600(725+x) = 118,300(470-x)$$

$$37,410,000 + 51,600x = 55,601,000 - 118,300x$$

$$169,900x = 18,191,000$$

$$x \approx 107$$

In Problem #5, many students could find an algebraic expression for the average speed:

$$\frac{d}{\frac{1}{2}d + \frac{1}{2}d},$$

where $d$ is the total distance, and $m$ and $n$ are the two speeds. This expression shows the process that most students use to calculate the average speed. It may be written in many forms. Writing fractions for division and omitting the dots for multiplication makes the expression easier to read and work with.

Algebra students may be challenged to simplify the expression:

$$\frac{d}{\frac{1}{2}d + \frac{1}{2}d} = \frac{d}{\frac{1}{2}dn + \frac{1}{2}dm} = \frac{d}{\frac{1}{2}dn + \frac{1}{2}dm} = \frac{d}{1} \cdot \frac{mn}{\frac{1}{2}dn + \frac{1}{2}dm}$$

$$= \frac{dmn}{\frac{1}{2}dn + \frac{1}{2}dm} = \frac{dmn}{\frac{1}{2}d(n+m)} = \frac{mn}{\frac{1}{2}(n+m)} = \frac{2mn}{n+m}$$

Encourage students to test the final expression! Notice that $d$ divides out when you simplify. This makes sense, because the answer does not depend on $d$!

Prealgebra students can try this, too! Even though it looks complicated, and they have not been taught algebraic procedures, they can *play* with the algebra: explore different expressions, test them to see if they work, make adjustments, and apply their knowledge of number properties to figure out how to work with variables. I am always impressed at how much progress students are able to make. And they learn that algebraic procedures make *sense*, because they come from something the students already know a lot about—numbers!

# Exploration 8

# Expanding and Contracting

This exploration is all about *similar* shapes. The mathematical meaning of *similar* is more exact than its everyday meaning of "about the same." Intuitively, similar shapes are what you get when you enlarge or reduce a figure (or leave it the same), but the angles, sides, and areas of similar shapes have precise mathematical relationships.

Your students may have run into similar shapes in earlier explorations! Lupe's ramps in Exploration 2 were similar when they had the same steepness. The different-sized triangles that students drew to help Derek and Beth find the height of the tower in Exploration 5 were also similar.

This example shows a few important terms: *corresponding, scale factor,* and *area factor.* I suggest introducing the word *corresponding* right away and waiting to discuss the other terms until they come up in Problem #1 or #2.

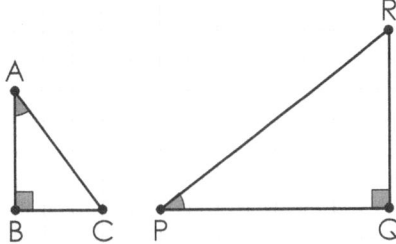

*Corresponding* sides or angles are in the same relative position on each figure. For example, $\overline{BC}$ and $\overline{QR}$ are corresponding sides, and $\angle A$ and $\angle P$ are corresponding angles. As you can see, you need to be careful when the figures are not facing the same direction!

The *scale factor* is the ratio of corresponding lengths. The scale factor from $\triangle ABC$ to $\triangle PQR$ is 2. It is the number by which you multiply $\triangle ABC$'s lengths to get the corresponding lengths for $\triangle PQR$. The *area factor* from $\triangle ABC$ to $\triangle PQR$ is $24 \div 6$ or 4, because the area of $\triangle PQR$ (24) is 4 times the area of $\triangle ABC$ (6).

*Note*: In earlier explorations, we usually thought of a ratio as two separate numbers (*a* of one thing for every *b* of something else). In this exploration, we also describe a ratio as a single number. For example, we say that a ratio of 6 to 3 equals 2, because 6 is 2 times greater than 3. This language is used in advanced courses and in many real-world situations.

DOI: 10.4324/9781003232797-11

# STAGE 1

Stage 1 is visual. By drawing a lot of pictures, your students will progress from a general knowledge of similar shapes to a detailed understanding of their properties. In order to make connections to earlier learning, encourage them to make tables and graphs of their measurements and to point out proportional relationships.

Problem #1 includes a trapezoid handout to give students larger shapes to work with. Do not worry if your students need a long time to solve the second trapezoid. It is challenging! In Problem #2, some students like to imagine the polygon as a character in a story or computer game. They can fill in the details!

A quick tip: Have students use the edge of a sheet of graph paper to measure the approximate length of a side when it is not vertical or horizontal.

## What You Will Need

» Graph paper.

## What Students Should Know

» Understand that similar figures have the same shape, but may be different sizes.
» Multiply and divide fractions.
» Measure area on a grid.

## What Students Will Learn

» Understand angle, length, and area relationships in similar figures.
» Solve challenging problems related to similar figures.
» Enhance spatial visualization skills.

# Problem #1

Level 1

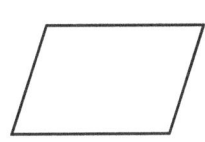 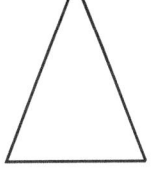 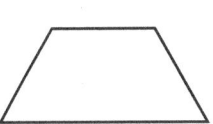

Level 2

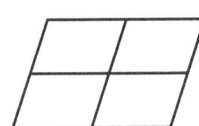 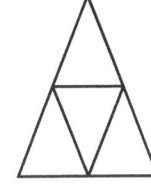 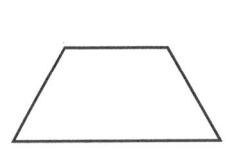

Level 3

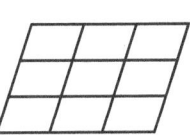 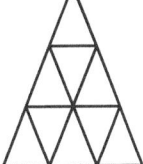 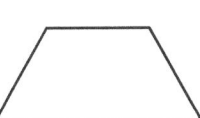

## Directions

- Fill in the trapezoids for Level 2 and Level 3.
- Draw a Level 4 picture for each figure.
- Make some observations and predictions.

### Testing the Waters

Solve the problem with triangles like these instead of the trapezoids.

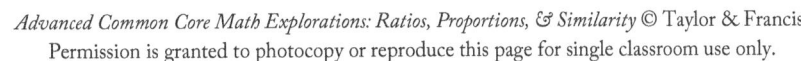

# TRAPEZOID HANDOUT
# FOR PROBLEM #1

# CONVERSATION STARTERS FOR #1

*What do you notice? What do you wonder?*

*I notice* that it is important to draw accurately.

*I wonder* how the side lengths of the inner shapes compare to the corresponding side lengths of the original shape at each level?

    This is an important question, both for conceptual understanding and for finding a practical way to draw the pictures!

*I notice* parallel sides in the parallelograms and triangles.

*I wonder* if the pattern of parallel sides will continue for the trapezoids?

*I notice* that I can experiment by making copies of the original shape and joining them.

    Pattern blocks may help you visualize what is happening!

*I notice* that I can decompose the trapezoid into small equilateral triangles.

*I notice* that the trapezoid can be built from three equilateral triangles.

*I notice* that an equilateral triangle can be built from three trapezoids.

*I notice* that I can use a Level 2 drawing to help me make a Level 4 drawing.

*I notice* that I can use the words *similar*, *ratio*, *scale factor*, and *area* in my observations.

*I wonder* what other shapes can be decomposed into similar shapes?

    Here's another one. Try to create some of your own!

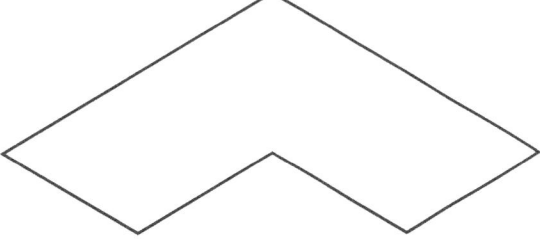

*I wonder* if the trapezoid (and other shapes) can be decomposed using scale factors of 5, 6, 7, etc.?

*I wonder* if there is an easy way to tell if a shape can be decomposed into similar shapes?

# SOLUTIONS FOR #1

*The Level 2 and Level 3 trapezoids:*

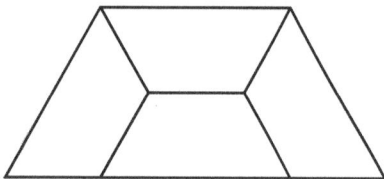 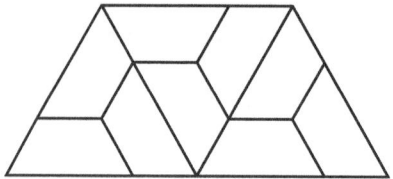

There are other solutions for the Level 3 trapezoid.

*The Level 4 figures:*

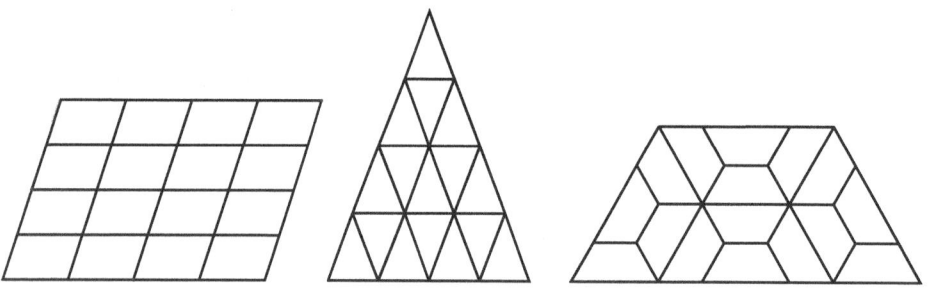

One strategy is to decompose the four parts from Level 2 into four parts each.

*Some observations:*

- » The inner polygons are similar to the original.
- » The scale factors from each inner polygon to the original polygon are 2, 3, and 4.
- » A square number of polygons fit into each original (4, 9, and 16).

*Some predictions:*

- » The next shapes in the patterns will decompose into 25 similar figures.
- » The number of inner polygons is always the square of the scale factor.
- » The area factor of similar shapes is always the square of the scale factor.

# Problem #2

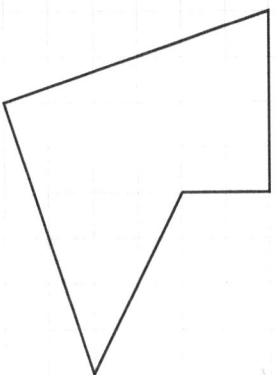

## Directions

- Create at least four polygons similar to the one above, two larger and two smaller.
- Investigate and describe ratios of angles, lengths, and areas between your polygons.
- Find a connection between the scale factor and area factor, and explain what causes it.

### Diving Deeper

Enlarge and reduce three-dimensional shapes. Explore the effects on surface area and volume.

### Testing the Waters

Do the problem with one of these polygons.

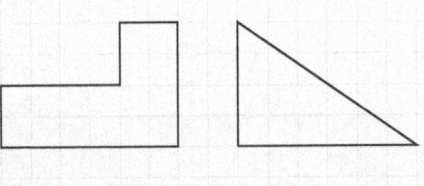

# CONVERSATION STARTERS FOR #2

*What do you notice? What do you wonder?*

*I wonder* how to draw the "diagonal" sides accurately?
> Pay attention to how far "over and up (or down)" you go to get from one endpoint to the other.

*I notice* that corresponding sides are parallel (unless I rotate or reflect the figure).

*I notice* that when I shrink the shape, the scale factor is less than 1.

*I notice* that the figure with a scale factor of 2 looks a lot larger than I expected.

*I notice* that I can use the edge of a piece of graph paper to measure lengths.

*I notice* that the lengths of my diagonal sides are not whole numbers.

*I notice* that I can decompose the shapes into rectangles and right triangles.

*I notice* that it is easier to write some scale factors as fractions than decimals.

*I notice* that scale factors in opposite directions are reciprocals.

*I notice* that the scale factor/area connection agrees with my observations in Problem #1.

*I wonder* what happens to each square unit when I enlarge the original figure?

*I wonder* what this tells me about the area of the enlarged figure?

*I wonder* if there are practical ways to create similar figures without graph paper?
> Does this picture give you any ideas?

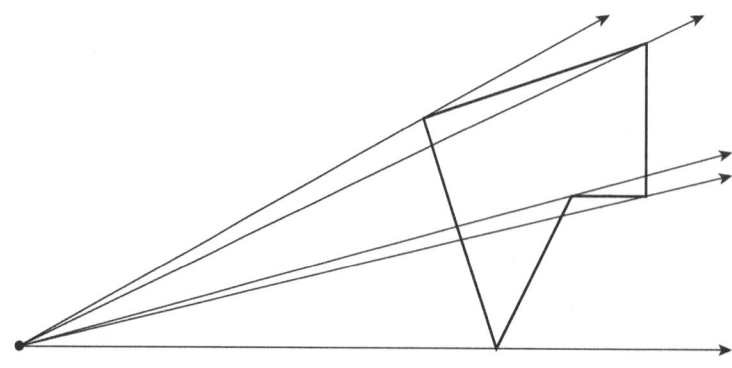

# SOLUTIONS FOR #2

*Some similar shapes with selected measurements*: Scale factors are in comparison to the original figure. Lengths are in units, and areas are in units². "~" indicates an approximate measurement.

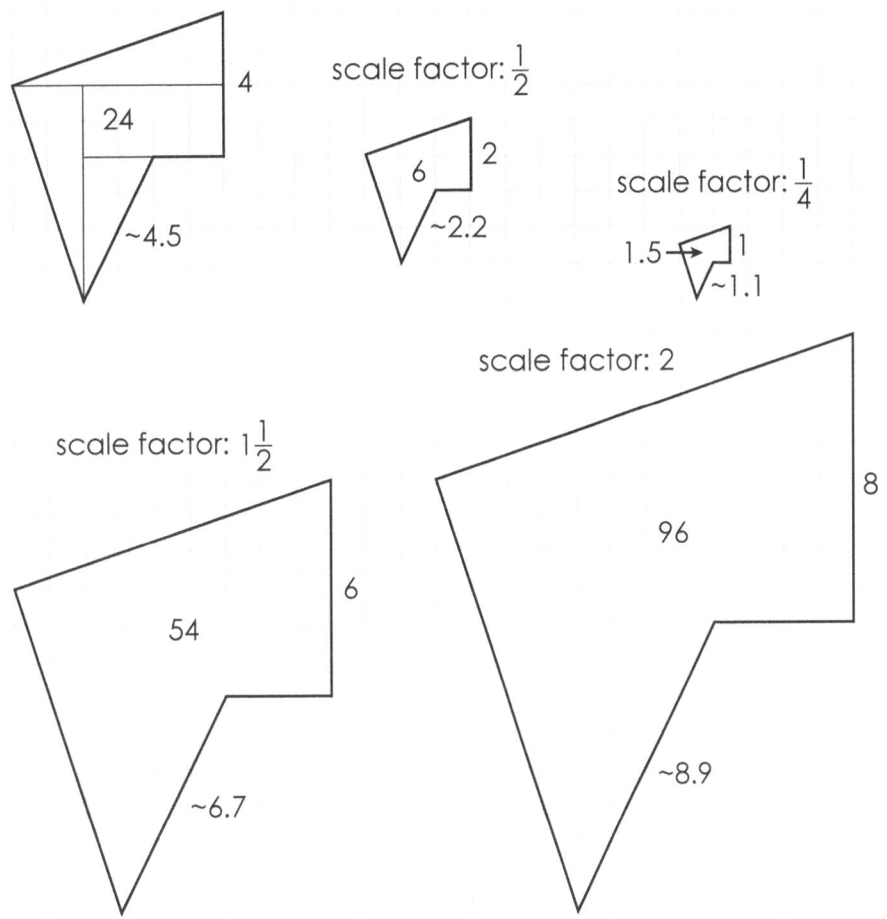

The original shape is decomposed is into simpler shapes to help find the area.

*Observations*: The ratio of corresponding angles of similar shapes is 1, because they are congruent. The area factor of two similar shapes is the square of the scale factor.

*Example:*

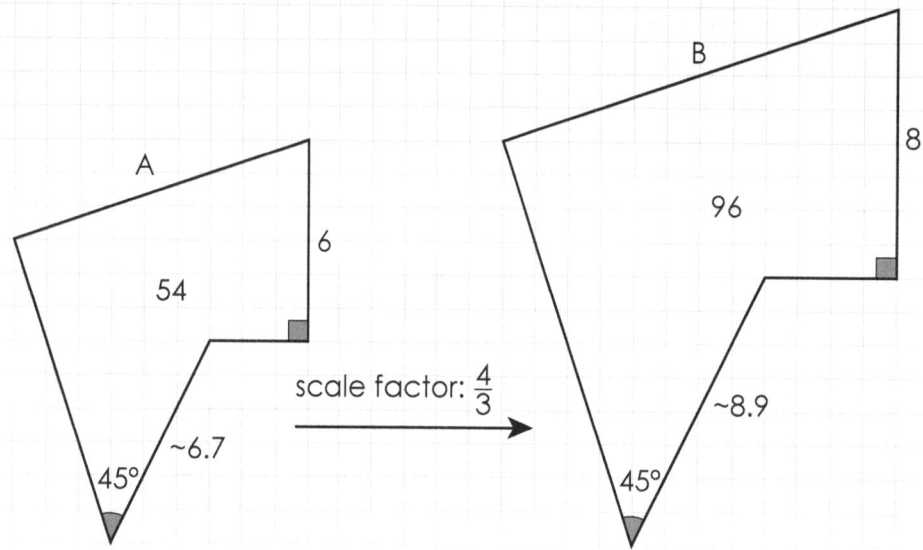

The scale factor from A to B is $\frac{4}{3}$, the ratio of corresponding side lengths:

$$8 \div 6 = \frac{8}{6} = \frac{4}{3} \text{ or } 8.9 \div 6.7 \approx \frac{4}{3}$$

The area factor from A to B is $96 \div 54 = \frac{96}{54} = \frac{16}{9}$, which is $\left(\frac{4}{3}\right)^2$.

Notice that the scale factor from B to A is $\frac{3}{4}$, the reciprocal of $\frac{4}{3}$!

*A reason for the scale factor/area factor connection:* Suppose the scale factor is $n$. Because the base and height of each square within a figure become $n$ times as large, the area of the square, and hence the entire figure, becomes $n^2$ times as large.

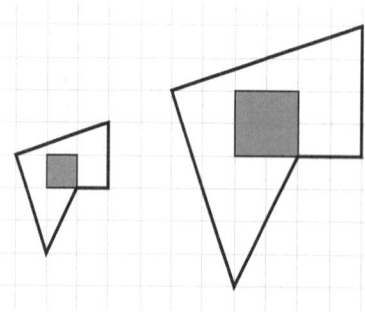

# STAGE 2

In Stage 2, students extend their understanding of similarity. In Problem #3, they combine Stage 1 knowledge with properties of polygons to find efficient ways to decide when polygons are similar. In Problem #4, students use similarity to solve a surprisingly challenging problem about right triangles. Those who know the Pythagorean theorem will be tempted to use it, but challenge them to find another way. Scale factors and area factors will do the trick, and using them will deepen students' understanding of similarity. As a bonus, this approach may enable some of them to prove the Pythagorean theorem algebraically! (See the Algebra Connections for more ideas.)

Before students begin Problem #4, they should have experience writing proportions to find missing lengths in similar figures until the ideas feel comfortable and familiar. Most prealgebra textbooks have plenty of exercises like these.

## *What Students Should Know*

- » Understand properties of squares, rectangles, parallelograms, and triangles.
- » Find the area of a triangle.
- » Corresponding angles of similar shapes are congruent.
- » Corresponding sides of similar shapes are in the same ratio.
- » The area factor in similar shapes is the square of the scale factor.

## *What Students Will Learn*

- » Make connections between properties of similarity and properties of polygons.
- » Reason with properties of polygons and similar shapes.
- » Recognize similar shapes within complex drawings.
- » Apply knowledge of similarity and proportions to solve challenging problems.

# Problem #3

You may be able to decide if two polygons are similar without measuring all of their sides and angles.

## Directions

Squares, Rectangles, Parallelograms, Triangles

- Find the number of measurements needed to prove similarity for each shape.
- Explain your thinking and justify all four of your answers.

# CONVERSATION STARTERS FOR #3

*What do you notice? What do you wonder?*

## For Squares

*I wonder* if I know in advance that both figures are squares?

> Yes. You may assume, for example, that all angles are 90° without measuring.

*I wonder* if there are shapes other than squares that are always similar?

> Think about other regular polygons!

## For Rectangles, Parallelograms, and Triangles

*I wonder* if it matters which angles or sides I measure?

> It often does, depending on the shape. Why?

*I wonder* which other angles I know in a parallelogram, once I know one of them?

> You know all of them. Why?

*I wonder* what happens to the sides when I change an angle in a triangle?

*I wonder* what happens to the angles when I change a side in a triangle?

*I wonder* if it helps to draw triangles inside of triangles?

*I wonder* how you can tell when two shapes with curved sides are similar?

> This requires a deeper understanding of similarity. However, most students will predict (correctly) that all circles are similar.

# SOLUTIONS FOR #3

*Squares*: 0 measurements are needed to prove that two squares are similar, because all squares are similar! Their corresponding angles are congruent, because every angle is a right angle. Pairs of corresponding sides have the same scale factor, because all sides are the same length.

*Rectangles*: 4 measurements are needed to prove that two rectangles are similar. You do not have to measure any angles, because they are all right angles. You need to measure two pairs of neighboring sides to ensure that they have the same scale factor.

*Parallelograms*: 6 measurements are needed to prove that two parallelograms are similar. You must measure one pair of corresponding angles to ensure that they are congruent. If so, the other pairs will automatically be congruent. (Why?) You need to measure two pairs of neighboring sides, just as you do for rectangles.

*Triangles*: Only 4 measurements are needed to prove that two triangles are similar! You must measure two pairs of corresponding angles. If each pair is congruent, the third will be, too. (Why?) When all pairs of corresponding angles are congruent, pairs of corresponding sides will automatically have the same scale factor! Encourage students to explore this by having them try to draw to nonsimilar triangles whose corresponding angles have the same measure. They will see that it is impossible! Ask them to describe their observations as they try to explain why. (They will learn to prove this fact in high school geometry once they have learned about the roles of definitions, axioms, and theorems in mathematical reasoning.)

# Problem #4

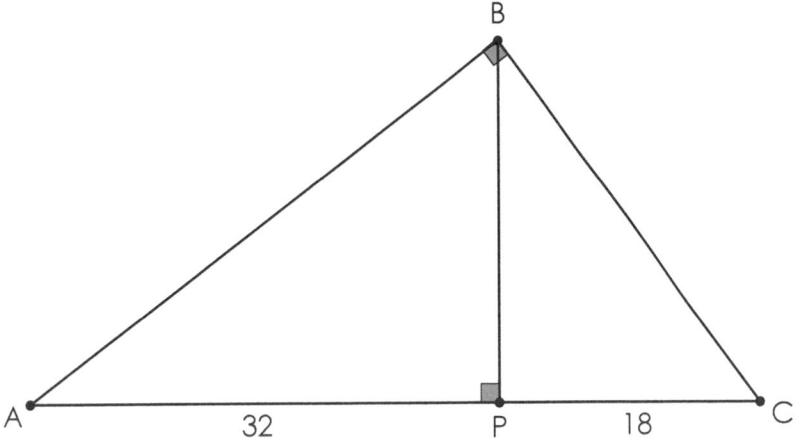

## Directions

- Find the area of $\triangle ABC$. Explain your thinking.
- Find any other measurements that you can. Explain your thinking.

### Diving Deeper

- Solve Problem #4 if the measurements are x and y instead of 32 and 18. (See Algebra Connections at the end of the exploration.)
- Suppose that $AB = a$, $BC = b$, and $AC = c$. Prove that $a^2 + b^2 = c^2$.

# CONVERSATION STARTERS FOR #4

*What do you wonder? What do you notice?*

*I notice* that there doesn't seem to be enough information at first.

*I notice* that it will help to know the height of $\triangle ABC$.

*I wonder* if the picture is drawn to scale?
>    Yes, it is. Feel free to use this to estimate other measurements. However, also find a way to determine exact values without measuring.

*I notice* that the drawing contains three right triangles.

*I notice* that all three triangles look similar!
>    How can you be sure that they are?

*I notice* that it is hard to visualize the corresponding sides.
>    Draw all three triangles separately and facing the same direction.

*I notice* that I can find areas of all the triangles once I find the height of the large triangle.

*I notice* that knowing the area relationships helps me find length relationships.

*I wonder* if it is possible to calculate the measures of the remaining angles?
>    Not quite yet! You need trigonometry to do this.

# SOLUTIONS FOR #4

$\triangle ABC$ has an area of 600 units$^2$.

*A strategy for finding the area*: Because you know the base, the challenge is to find the height. The three similar triangles in the figure make this possible.

*Why the three triangles are similar*: All three triangles have a right angle, and both small triangles have another angle in common with the large triangle. Based on Problem #3, the two small triangles are similar to the large triangle—and because they are both similar to the same triangle, they must also be similar to each other.

*Redrawing the similar triangles to make it easier to see corresponding sides*:

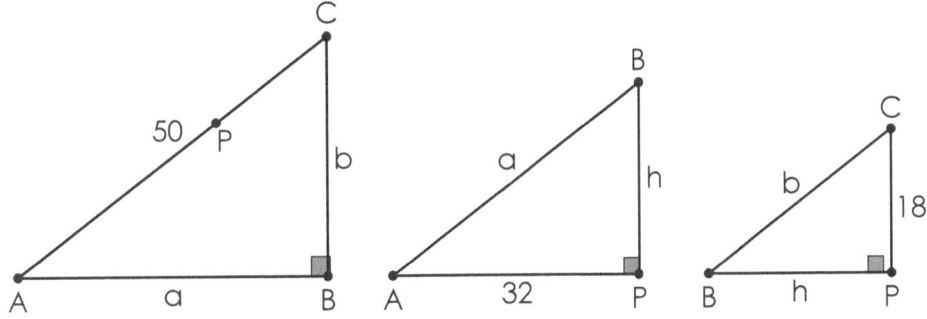

*Calculating the height, h*: Because corresponding sides have the same ratio, $\dfrac{32}{h} = \dfrac{h}{18}$. Using cross products:

$$h \cdot h = 32 \cdot 18 \qquad h^2 = 576 \qquad h = \sqrt{576} = 24$$

*Calculating the area of $\triangle ABC$*:

$$A = \frac{1}{2} \cdot 50 \cdot 24 = 600 \ \text{units}^2$$

*Calculating the other measurements*: Now that you know $h$, you can calculate the areas of the other two triangles.

$$\triangle APB: \ A = \frac{1}{2} \cdot 32 \cdot 24 = 384 \ \text{units}^2$$

$$\triangle BPC: \ A = \frac{1}{2} \cdot 24 \cdot 18 = 216 \ \text{units}^2$$

*Area factors and scale factors enable you to calculate a and b. For example*:

» The area factor for $\triangle ABC$ to $\triangle APB$ is $384 \div 600 = \dfrac{384}{600} = \dfrac{16}{25}$.

» The scale factor for $\triangle ABC$ to $\triangle APB$ is $\sqrt{\dfrac{16}{25}} = \dfrac{4}{5}$.

» The value of $a$ is $50 \cdot \dfrac{4}{5} = 40$.

You can use the same strategy on $\triangle ABC$ and $\triangle BPC$ to find that $b = 30$.

# STAGE 3

In Stage 3, students explore two famous mathematical topics related to ratios and similarity: the Fibonacci sequence and the golden ratio. The Fibonacci sequence is the infinite list of numbers $1, 1, 2, 3, 5, 8, 13, 21 \ldots$, in which each number is the sum of the two numbers before it. It appeared around the year 1200 in a problem posed by Leonardo Pisano, also known as Fibonacci. Its popularity has to do with its countless intriguing patterns, its tantalizing appearances in the natural world (in pine cones and flowers, for example), and its connections to the golden ratio.

The golden ratio, $\varphi$ (phi), like $\pi$, is an *irrational* number, meaning that it cannot be written as a simple fraction (with a whole number numerator and denominator), and that its decimal never terminates (stops) or repeats. It is approximately equal to 1.618033989. People have found connections to the golden ratio in art, architecture, biology, music, and elsewhere. As students will learn, you can use ratios of numbers in the Fibonacci sequence to approximate the golden ratio!

Just one tip: In Problem #6, allow students plenty of time to look at the diagram. It is complex, but they may be able to figure out what is happening on their own!

## What Students Should Know

&raquo; Understand ratios and similarity (at the level of Stage 2).

&raquo; Understand that the reciprocal of $n$ is $1 \div n$, because the product of reciprocals is 1.

## What Students Will Learn

&raquo; Analyze and extend complex patterns.

&raquo; Apply ratios and similarity to solve challenging problems.

# Problem #5

1, 1, 2, 3, 5, 8, 13, . . . ?

## Directions

- Continue the pattern. Explain your thinking.
- Find and describe more patterns in the list.

### Diving Deeper

Begin the list with the numbers 1, 3 instead of 1, 1. Do your patterns still work?

# CONVERSATION STARTERS FOR #5

*What do you notice? What do you wonder?*

*I wonder* what happens when I find differences of consecutive numbers?
> You get the list back again!

*I wonder* how long the list continues?
> It goes on forever.

*I notice* that there is a pattern of odd and even numbers.

*I notice* patterns in factors of numbers in the list.

*I notice* that interesting things happen when I add numbers in the list.

*I notice* that interesting things happen when I square numbers in the list.

# SOLUTIONS FOR #5

*The list*:

$$1, 1, 2, 3, 5, 8, 13, 21, 34, 55, 89, 144, 233, 377, 610, \ldots$$

Each number is the sum of the two preceding numbers. This is the famous *Fibonacci sequence*.

*More patterns in the Fibonacci sequence*:

Let's call the *n*th Fibonacci number $F_n$. For example, to say that the sixth Fibonacci number is 8, we write $F_6 = 8$. Here are just a few of many amazing patterns!

» The Fibonacci numbers follow the pattern *odd, odd, even, odd, odd, even*, etc.

» Every third Fibonacci number is a multiple of 2; every fourth number is a multiple of 3; every fifth number is a multiple of 5; and every 12th number is a multiple of 6. (How does this continue?)

» The sum of the first *n* Fibonacci numbers is one less than the $(n+2)$ nd Fibonacci number. In symbols: $F_1 + F_2 + \cdots + F_n = F_{n+2} - 1$.

» $F_n^2 + F_{n+1}^2 = F_{n+(n+1)}$. For example $F_3^2 + F_4^2 = F_7$ (i.e., $2^2 + 3^2 = 13$).

» $F_n^2$ is alternately one more and one less than $F_{n-1} \cdot F_{n+1}$. For example, $5^2$ is one more than $3 \cdot 8$, and $8^2$ is one less than $5 \cdot 13$.

Students may discover many other patterns. We will investigate patterns related to ratios and similarity in the next two problems.

# Problem #6

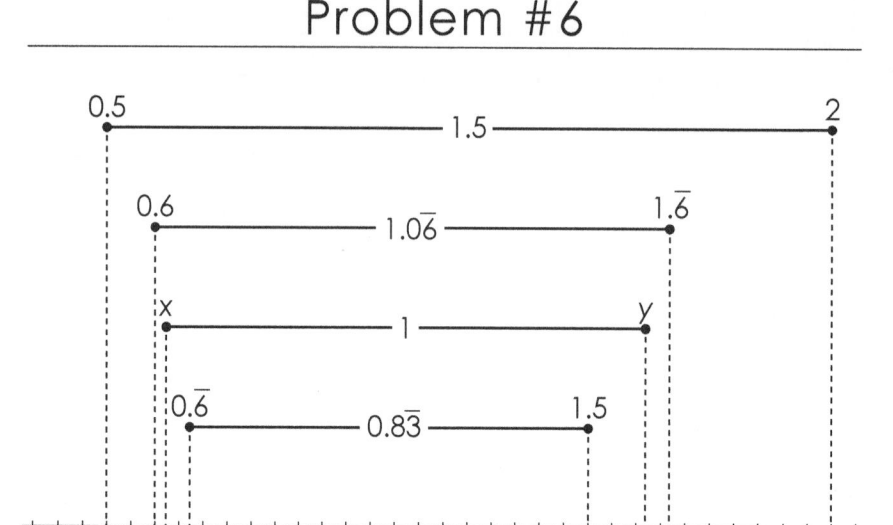

## Directions

- Find the values of x and y. Explain your thinking.
- Describe a connection to Problem #5.

## Diving Deeper

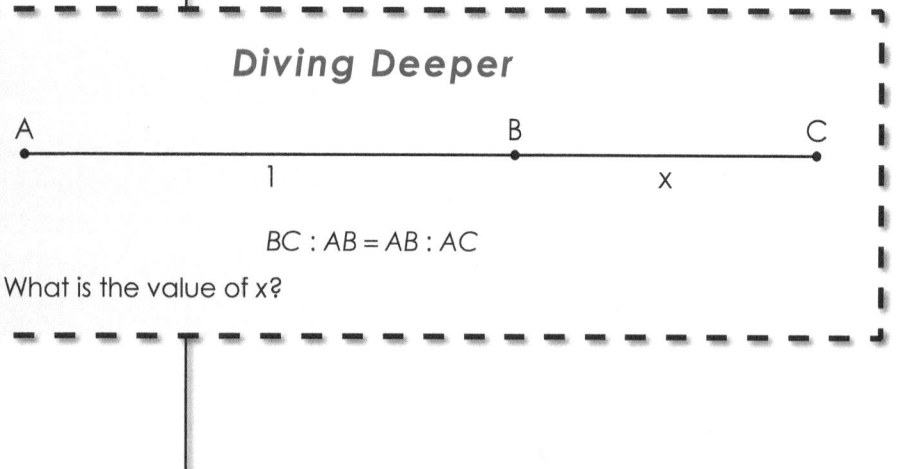

$$BC : AB = AB : AC$$

What is the value of x?

# CONVERSATION STARTERS FOR #6

*What do you notice? What do you wonder?*

Remember to give students plenty of time to analyze the diagram without your help before giving them too much direct guidance! There is a lot going on, but they can discover most of it themselves if they are persistent. Encouraging them to collaborate may help.

*I notice* that the vertical bars are getting closer together.

*I wonder* where the vertical bars meet as they move inward?
> They meet at the number 1.

*I wonder* what the number between the vertical bars means.
> It is the distance between the bars.

*I wonder* what the connection is between the numbers in each pair marked by the bars?
> The connection may be easier to see if you multiply the two numbers in each pair, or write each decimal as a fraction.

*I wonder* how to find the reciprocal of a decimal when I don't know its fraction?

## If Students Decide To Try a Guess/Test Process

*I wonder* if I should use fractions or decimals?

*I wonder* whether I should increase or decrease my guess?

*I wonder* how much to increase or decrease my guess?

*I wonder* how many decimal places I should keep track of?

*I wonder* what is the best way to organize my work?

*I wonder* what the Fibonacci numbers have to do with this problem?
> Look at the fraction forms of the decimals.

# SOLUTIONS FOR #6

*The values of x and y*: The pairs of vertical bars show reciprocals. *x* and *y* are the positive reciprocals that have a difference of 1:

$$x \approx 0.61803398875 \qquad y \approx 1.61803398875$$

The value of *y* is known as the Golden Ratio (written $\varphi$ [phi]), a famous *irrational* number connected to the Fibonacci sequence.

*A strategy for finding x and y*: Using the number line, make a guess of about 0.61 for *x*. The reciprocal of 0.61 is $1 \div 0.61 \approx 1.6393442623$. The difference is greater than 1. The number line shows that you must increase your guess to decrease the difference.

| *x* (guess) | *y* (reciprocal) | difference |
|---|---|---|
| 0.61 | 1.6393442623 | 1.0293442623 |
| 0.62 | 1.61290322581 | 0.992903225807 |
| 0.617 | 1.62074554295 | 1.00374554295 |

Continue guessing and testing for as long as you like. Your answer will never be perfect (because $\varphi$ is irrational), but it will get closer and closer to $\varphi$.

*Note*: Some students may choose to work with fractions instead of decimals.

*A connection to Problem #5*: If you write the numbers as fractions, the numerators and denominators are Fibonacci numbers! Ratios of consecutive pairs in the Fibonacci sequence (and their reciprocals) get closer to *x* and *y* as you use larger numbers in the sequence.

$$\frac{1}{2} = 0.5 \qquad\qquad \frac{2}{1} = 2$$

$$\frac{2}{3} = 0.\overline{6} \qquad\qquad \frac{3}{2} = 1.5$$

$$\frac{3}{5} = 0.6 \qquad\qquad \frac{5}{3} = 1.\overline{6}$$

$$\frac{5}{8} = 0.625 \qquad\qquad \frac{8}{5} = 1.6$$

$$\frac{8}{13} = 0.\overline{615384} \qquad \frac{13}{8} = 1.625$$

$$\frac{13}{21} = 0.\overline{619047} \qquad \frac{21}{13} = 1.\overline{615384}$$

There are some great patterns here! The difference between the reciprocals keeps getting closer to 1. The digits at the beginning of the decimals eventually stop changing. The improper fraction on the right is alternately greater and less than the Golden Ratio, always getting closer to it.

*I wonder* what happens with the ratio of every second or third Fibonacci number?

*I wonder* what happens if you begin the sequence with 1, 3 instead of 1, 1?

# Problem #7

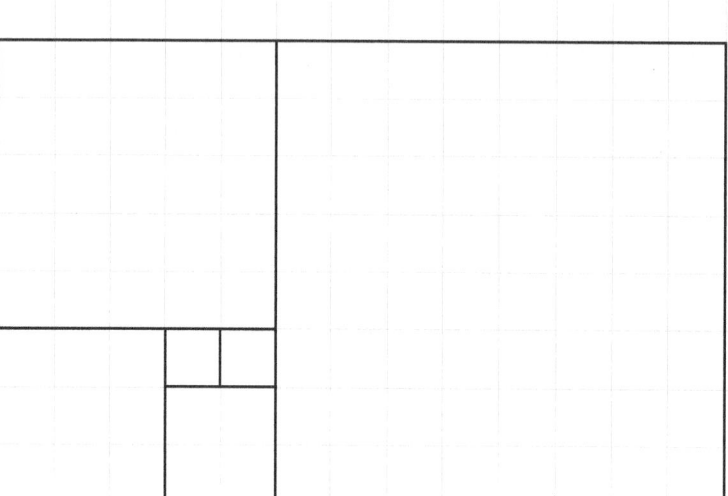

## Directions

- Discover the pattern and continue the drawing. Explain your thinking.
- Investigate the (nonsquare) rectangles. Are they similar?
- Explain what happens to the rectangles as the picture gets larger.
- Explore the areas of the rectangles and their ratios.

## Diving Deeper

$$1^2 + 1^2 + 2^2 + 3^2 = 3 \cdot 5$$

Use the drawing to create more equations like this one. Describe the pattern.

# CONVERSATION STARTERS FOR #7

*What do you notice? What do you wonder?*

*I notice* a pattern in the side lengths of the squares.

*I notice* that the squares get larger in a counterclockwise pattern.

*I notice* length ratios that remind me of Problem #6.

*I notice a surprising pattern:*

$$\frac{1}{\varphi} \approx 0.618033989 \qquad \varphi \approx 1.618033989 \qquad \varphi^2 \approx 2.618033989$$

*I wonder* if $\varphi^3 \approx 3.618033989$ ?
> You can estimate to answer this question!

*I wonder* what would happen if I could draw the diagram in reverse (working from the outside in) beginning with a "Golden Rectangle" having a base to height ratio of $\varphi : 1$ ?
> There would be one square of each size, and the pattern would continue spiraling inward forever! All of the rectangles would be exactly similar.

*I wonder* how this spiral is created?

# SOLUTIONS FOR #7

*Extending the pattern*:

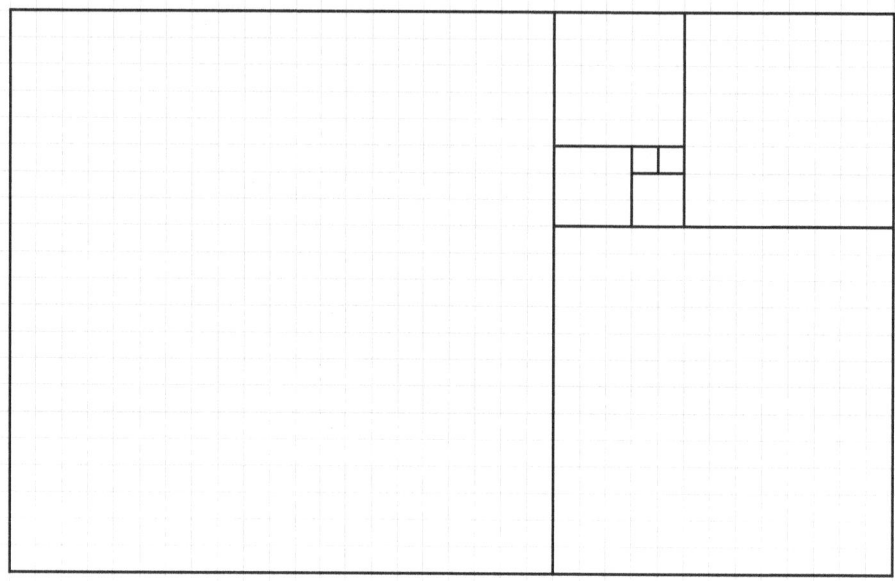

This picture shows two more steps in the pattern. The next step would be to draw a 34 by 34 square above this picture.

*Describing the pattern*: The side lengths of the squares come from the Fibonacci sequence. Start with a 1 by 1 square. New squares are attached to the figure according to the pattern right, below, left, above, right, below, left, above, etc.

*Relationships between the rectangles*:

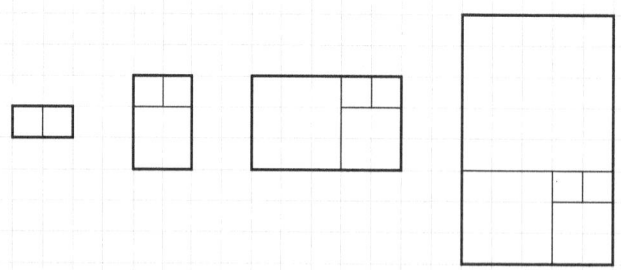

The rectangles are not similar. However, as they get larger, consecutive rectangles get closer to being similar. You can see this two ways:
  » The ratio between a pair of shorter sides is not the same as the ratio between the longer sides. However, both ratios get closer to each other (and to $\varphi$!) as the rectangles get larger.

» The ratio of the longer side to the shorter side *within* a rectangle also gets closer to $\varphi$ as the rectangles get larger. The long : short ratios from left to right are $\frac{2}{1}, \frac{3}{2}, \frac{5}{3}, \frac{8}{5}, \frac{13}{8}$, etc. These ratios contain the Fibonacci numbers, and as we saw in Problem #6, they keep getting closer to the Golden Ratio, $\varphi$.

*The areas of the rectangles*: The areas of the rectangles are 2, 6, 15, 40, 104, 273, etc. Their ratios are:

$$6 \div 2 = 3 \quad 15 \div 6 = 2.5 \quad 40 \div 15 = 2.\overline{6} \quad 104 \div 40 = 2.6 \quad 273 \div 104 = 2.625$$

Because the rectangles are almost similar, the area factor is close to the square of the length ratios. In other words, the ratio of the areas of consecutive rectangles is getting closer to $\varphi^2$ (approximately 2.618033989).

# ALGEBRA CONNECTIONS

The Diving Deeper questions in Problem #4 ask students to explore the situation algebraically, replacing the numbers 32 and 18 by the variables $x$ and $y$.

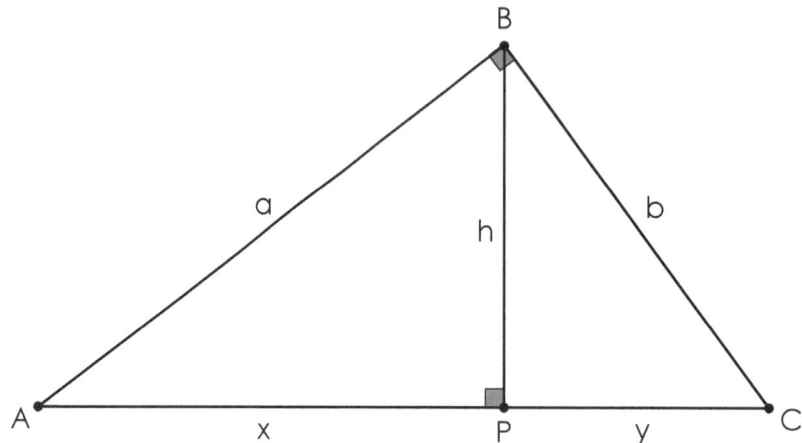

Prealgebra students may be able to make progress on the first question! First, they may relabel the picture of the individual triangles.

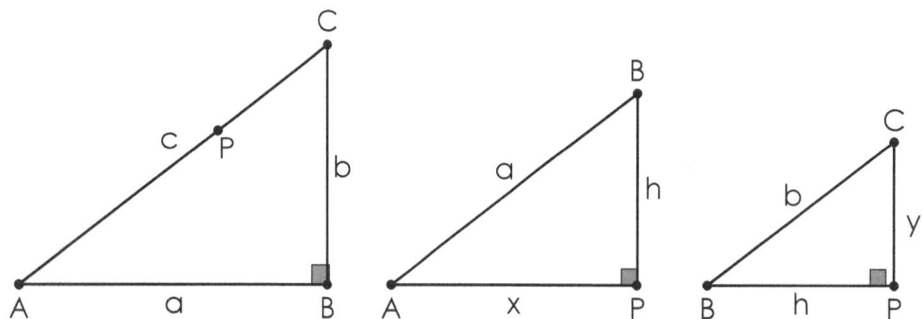

From the two triangles on the right, they may write the proportion $\dfrac{x}{h} = \dfrac{h}{y}$. Using cross products gives $h^2 = xy$ or $h = \sqrt{xy}$. (This is the *geometric mean* of $x$ and $y$.) Using this expression for the height, they can find expressions for the areas of all three triangles:

$$\triangle ABC: \frac{1}{2}(x+y)\sqrt{xy} \qquad \triangle APB: \frac{1}{2}x\sqrt{xy} \qquad \triangle BPC: \frac{1}{2}y\sqrt{xy}$$

Students may test these expressions by substituting the values from Problem #4!

Algebra students (and adventurous prealgebra students) may be able to go further. Following the same process as they did with the numbers, they can use the area factors and scale factors to find expressions for the lengths $a$ and $b$. With some simplification, the answers may be written:

$$a = \sqrt{x(x+y)} \text{ and } b = \sqrt{y(x+y)}$$

From these expressions and the fact that $x+y=c$, they can prove the Pythagorean theorem!

$$a^2 + b^2 = x(x+y) + y(x+y) = (x+y)(x+y) = (x+y)^2 = c^2$$

In Problem #6, algebra students who know the quadratic formula may use it to calculate the exact value of $\varphi$. Begin by writing a formula stating that the difference of the reciprocals is 1:

$$x - \frac{1}{x} = 1$$

Multiply both sides by $x$. Then subtract $x$ from both sides:

$$x^2 - 1 = x \qquad x^2 - x - 1 = 0$$

Complete the square or apply the quadratic formula and choose the positive solution to obtain:

$$x = \frac{1 + \sqrt{5}}{2},$$

which is the exact value of $\varphi$!

# Exploration 9

## Pythagorean Connections

In this exploration, students connect the Pythagorean theorem to similarity and proportion concepts in order to make new discoveries about square roots and to solve a challenging measurement problem. Before they begin, check that they know the words *congruent* ($\cong$) and *similar* (~), and remind them that corresponding vertices belong in the same position of polygons' names. For example, $\triangle MTC \sim \triangle AXR$ means that the triangles are similar, and that $M$ and $A$, $T$ and $X$, and $C$ and $R$ respectively are in corresponding positions in the triangles.

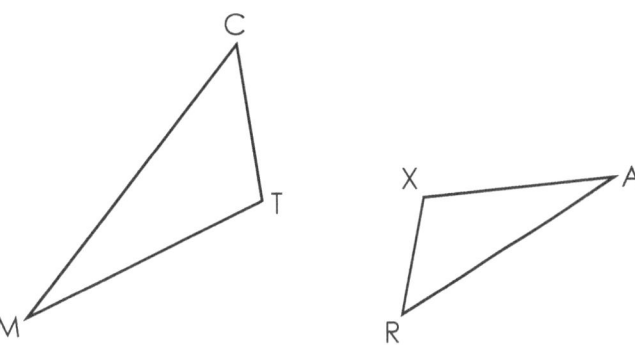

The Pythagorean theorem belongs to the eighth-grade standards in the Common Core. In my experience, it fits well into talented students' learning progression by sixth grade or so if they have a strong understanding of area and square roots. For a conceptual introduction to the Pythagorean theorem, see the exploration "A New Slant on Measurement" from the Measurement and Polygons book in this series.

DOI: 10.4324/9781003232797-12

# STAGE 1

In Stage 1, students use similar shapes and the Pythagorean theorem to make new discoveries about square roots. This will give them a strong background for simplifying square roots in algebra.

To prepare students for Problem #1, introduce the words *radical* and *radicand*. A radical is a root. This includes square roots, cube roots, fourth roots, etc. (The focus of this problem is on square roots.) The radical symbol is "$\sqrt{\phantom{x}}$." The *radicand* is the expression inside the radical symbol. For example, the radicand of $\sqrt{7}$ is 7.

Before beginning, explain the difference between *exact* and *approximate* forms of radicals. To write a radical in *exact* form, express it with the radical symbol. To write it in *approximate* form, show it as a rounded decimal. For example, an exact form of the square root of 20 is $\sqrt{20}$, and an approximate form is 4.47.

## What Students Should Know

>   » Understand square roots.
>   » Understand and use the Pythagorean theorem.
>   » Understand ratios, proportions, and similarity.

## What Students Will Learn

>   » Analyze patterns and make predictions in similar triangles.
>   » Find equivalent expressions for square roots.

# Problem #1

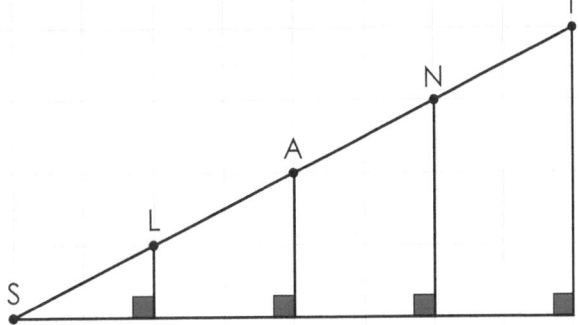

## Directions

- Name the exact lengths of $\overline{SL}$, $\overline{SA}$, $\overline{SN}$, and $\overline{ST}$ in at least two ways.
- Compare different ways of naming the length of a segment. Describe patterns and make observations.
- Find another way to write $\sqrt{52}$ in exact form. Explain your thinking and draw a picture to justify your answer.

### Diving Deeper

- Explain why $\sqrt{160} - \sqrt{90} = \sqrt{10}$. Draw a picture to support your answer.
- Create a rule for writing square roots in exact form in more than one way. Test your rule on some square roots and explain why it works.

# CONVERSATION STARTERS FOR #1

*What do you notice? What do you wonder?*

*I notice* four right triangles in the drawing.

*I notice* that all four triangles share the same angle.

*I notice* that all four triangles are similar.

*I notice* that I can use the Pythagorean theorem to find lengths of slanted sides.

*I notice* a pattern in the radicands of the lengths.

*I notice* that I can also use similar triangles to find lengths of slanted sides.

*I notice* that I can use a calculator to check that expressions appear to be equal.

*I notice* that 52 has a factor that is a square number (4) and that $52 \div 4 = 13$ .

*I notice* that 13 is a sum of two square numbers.
  This helps with drawing a triangle whose hypotenuse has a length of $\sqrt{13}$ .

*I notice* that I can create more similar triangles in the picture (an infinite number).

*I notice* trapezoids in the drawing.

*I notice* that I can use triangles to find side lengths and areas of some trapezoids.

*I wonder* how to make a right triangle having a hypotenuse of length $\sqrt{3}$ or $\sqrt{6}$ ?
  Suppose that the lengths of the legs are not whole numbers.

*I wonder* if I can use sums of *three* square numbers to create drawings like the ones in this problem?
  Try drawing some three dimensional pictures. For example, think of a diagonal through a cube.

# SOLUTIONS FOR #1

*Finding lengths using the Pythagorean theorem and similarity:*

| Segment | Scale factor | Similarity | Pythagorean theorem | Approximation |
|---|---|---|---|---|
| $\overline{SL}$ | 1 | $\sqrt{5}$ | $\sqrt{2^2+1^2}=\sqrt{5}$ | 2.236 |
| $\overline{SA}$ | 2 | $2\cdot\sqrt{5}$ | $\sqrt{4^2+2^2}=\sqrt{20}$ | 4.472 |
| $\overline{SN}$ | 3 | $3\cdot\sqrt{5}$ | $\sqrt{6^2+3^2}=\sqrt{45}$ | 6.708 |
| $\overline{ST}$ | 4 | $4\cdot\sqrt{5}$ | $\sqrt{8^2+4^2}=\sqrt{80}$ | 8.944 |

*Patterns and observations*: Because the triangles are similar, the length of each slanted segment is equal to the original length, $\sqrt{5}$, multiplied by the scale factor from the smallest triangle. In the Pythagorean theorem result, the radicand is equal to the square of the scale factor multiplied by 5. You can use a calculator to show that both forms of the radical for each length appear to be equal: for example, $\sqrt{20}=2\cdot\sqrt{5}\approx4.472$.

*Another way to write* $\sqrt{52}$: Based on the patterns above, look for a perfect square factor of 52. Because $13\cdot4=52$ and $4=2^2$, create a right triangle with a hypotenuse of $\sqrt{13}$, and then make a similar triangle using a scale factor of 2.

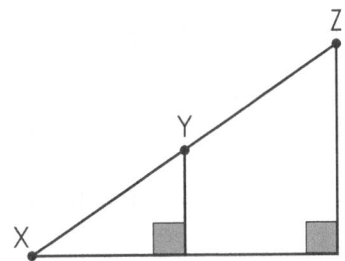

Length of $\overline{XY}$: $\sqrt{3^2+2^2}=\sqrt{13}$      Length of $\overline{XZ}$: $\sqrt{6^2+4^2}=\sqrt{52}$

Because $\overline{XZ}$ is twice the length of $\overline{XY}$, $\sqrt{52}=2\cdot\sqrt{13}$.

# STAGE 2

Problem #2 is a challenging task based on a simple picture. Try giving it to students without the directions. Let them look at the picture and think of their own questions to explore. If they do not suggest the question in the directions, guide them to it, but encourage them to follow up on their own questions later!

Most students solve the problem using the Pythagorean theorem. An approach using similarity is discussed in Stage 3. If students discover the Stage 3 method first (which is rare), suggest that they use the Pythagorean theorem to check their answer.

Before beginning, clarify the meaning of *GP* and *PH*. A bar over two letters indicates a segment. When the bar is not there, it stands for the *length* of the segment. For example, $\overline{GP}$ is a segment, and *GP* is its length.

It is fun to ask students to create a story to fit this problem. I originally heard the problem as a story about a camper (*G*) who needs to stop by a river ($\overline{LM}$) before returning to the campfire (*H*). Maybe your students can create a more realistic (or more entertaining) story!

## What Students Should Know

» Understand and use square roots and the Pythagorean theorem.

## What Students Will Learn

» Solve a challenging problem using the Pythagorean theorem.
» Organize data and apply a guess/test strategy for a complex problem.

# Problem #2

$P$ slides along $\overline{LM}$. $G$, $H$, $L$, and $M$ do not move.

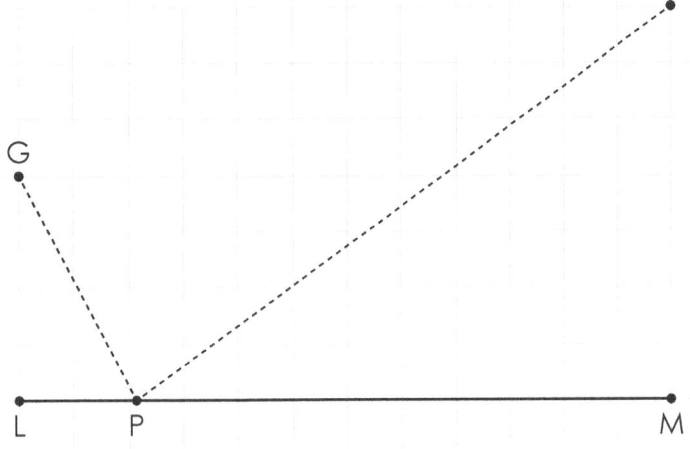

## Directions

- Make $GP + PH$ as small as possible.
- Show the location of $P$ and the total length, $GP + PH$, that solve the problem. Explain your strategy.

## Diving Deeper

- Create a story for this math problem.
- Think of other questions for the picture and try to answer them.

# CONVERSATION STARTERS FOR #2

*What do you notice? What do you wonder?*

*I wonder* if $P$ will be closer to $L$ or to $M$?

*I notice* that $\overline{GL}$ and $\overline{HM}$ make right angles with $\overline{LM}$.

*I wonder* how many decimal places I should use in my calculations?

*I notice* that $GP + PH$ does not change at a constant rate compared to $LP$.

*I notice* that $GP + PH$ changes very slowly near its minimum.

*I wonder* how I can tell when I have found the exact minimum?

## Other Questions

*I wonder* if the two triangles are similar when $GP + PH$ is a minimum?
They are! It will be easier to see why after Stage 2.

*I wonder* if anything else special happens when $GP + PH$ is a minimum?

*I wonder* what location(s) for $P$ makes $GP$ equal to $PH$?

*I wonder* what location(s) for $P$ makes $\angle GPH$ a right angle?

*I wonder* what location(s) for $P$ makes $\angle GPH$ as large (small) as possible?

*I wonder* what location(s) for $P$ makes $\angle GPL$ and $\angle HPM$ congruent?

*I wonder* what location(s) for $P$ makes the area of $\triangle GPH$ as large (small) as possible?

*I wonder* if any of these things happen at the same point?

Dynamic geometry software such as GeoGebra® is a great tool for exploring some of these questions.

# SOLUTIONS FOR #2

*A guess/test strategy*: Show the right triangles $\triangle GLP$ and $\triangle HMP$. Make guesses for the length $LP$, and use the Pythagorean theorem to calculate $GP + PH$.

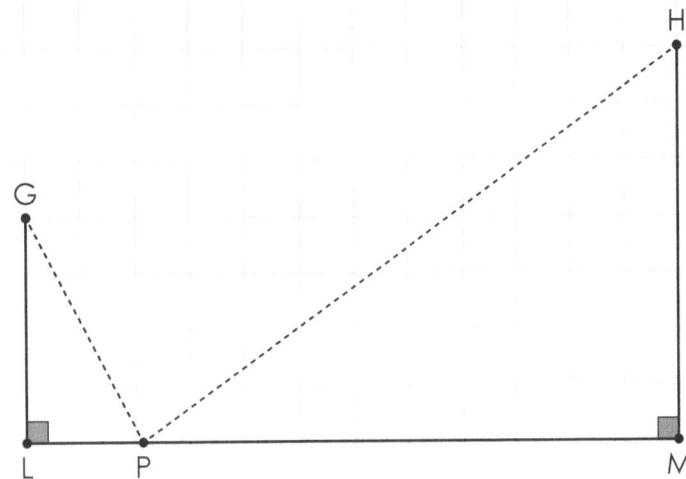

It is easier to see patterns when you organize the work:

| LP (guess) | GP | PH | GP + PH |
|------------|------------------------------|-------------------------------|---------|
| 0 | 4 | $\sqrt{12^2 + 7^2} \approx 13.89$ | 17.89 |
| 1 | $\sqrt{4^2 + 1^2} \approx 4.12$ | $\sqrt{11^2 + 7^2} \approx 13.04$ | 17.16 |
| 2 | $\sqrt{4^2 + 2^2} \approx 4.47$ | $\sqrt{10^2 + 7^2} \approx 12.21$ | 16.68 |
| 3 | $\sqrt{4^2 + 3^2} = 5$ | $\sqrt{9^2 + 7^2} \approx 11.40$ | 16.40 |
| 4 | $\sqrt{4^2 + 4^2} \approx 5.66$ | $\sqrt{8^2 + 7^2} \approx 10.63$ | 16.29 |
| 5 | $\sqrt{4^2 + 5^2} \approx 6.40$ | $\sqrt{7^2 + 7^2} \approx 9.90$ | 16.30 |

$GP + PH$ "turns around" near 4. If students continue to explore values near 4, they will find that the minimum length happens when $P$ is about 4.36 units to the right of $L$. Some students may investigate values of $LP$ greater than 5 if they think a smaller total length may be hiding there.

# STAGE 3

In Stage 3, students use reflections, similar triangles, and proportions to solve the problem from Stage 2 in a more efficient way. I usually start by suggesting that students experiment with applying different transformations in the picture. When I give them enough time to explore, someone usually discovers the necessary reflection. If not, I suggest that they reflect $G$ over $\overline{LM}$ and let them take it from there.

Begin by reminding students about the vocabulary: translations (slides), reflections (flips), and rotations (turns). These are the transformations they should experiment with. To complete the problem, it will help if students know how to solve simple linear equations. If not, they may use a guess and test strategy to solve the final proportion.

## What Students Should Know

» Translate, rotate, and reflect geometric figures.
» Understand and apply ratios, proportions, and similarity.
» Solve linear equations (recommended).

## What Students Will Learn

» Recognize similar figures within a larger drawing.
» Apply similarity and proportions to solve a challenging problem.

# Problem #3

$P$ slides along $\overline{LM}$. $G$, $H$, $L$, and $M$ do not move.

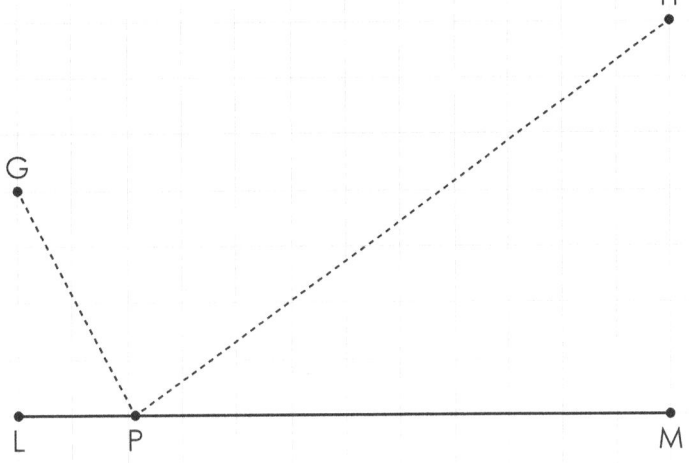

Sometimes, you can use geometric transformations to understand problems in new ways.

## Directions

- Use a transformation to help you make $GP + PH$ as small as possible.
- Calculate a number for the exact location of $P$ that solves the problem.
- Explain your thinking with words, drawings, and calculations.

### Diving Deeper

Create a formula for the location of $P$ that makes $GP + PH$ as small as possible. (What variables should you use?)

# CONVERSATION STARTERS FOR #3

*What do you notice? What do you wonder?*

*I wonder* which transformation I should try?

*I wonder* which parts of the picture I should transform?

There are not too many choices: translations, rotations, and reflections of a few points or segments. Don't be afraid to dig in and try things that might not work!

*I notice* that the shortest distance between two points is along a straight path.

## After Students Have Located P On The Drawing (See The Solutions)

*I notice* that many angles in the picture are congruent.

Pay special attention to $\angle LPF$ and $\angle HPM$. They are congruent because they are *vertical* angles.

*I notice* some similar triangles.

Be sure you can explain how you know they are similar.

*I notice* a lot of unknown lengths in the similar triangles.

Focus on the lengths you know and those that will help you find a number for the location of *P*.

*I notice* that the similar triangles are facing in different directions.

Yes, you have to be careful when you identify corresponding angles and sides.

## As Students Are Thinking Of Writing An Equation

*I notice* that it may help to use a variable.

*I notice* that my equation has two different variables.

That makes it harder to solve.

*I notice* that there is a predictable relationship between *LP* and *PM*.

This may help you write an equation using only one variable.

# SOLUTIONS FOR #3

*A strategy using reflections and similarity:* Reflect $G$ across $\overline{LM}$ to $F$. Because $FP$ and $GP$ are equal, $GP + PH = FH$. To make the total length as small as possible, you just need to make a straight path from $F$ to $H$. You can draw the correct location for $P$ without doing a single calculation! Its location on $\overline{LM}$ seems to match the result in Stage 2.

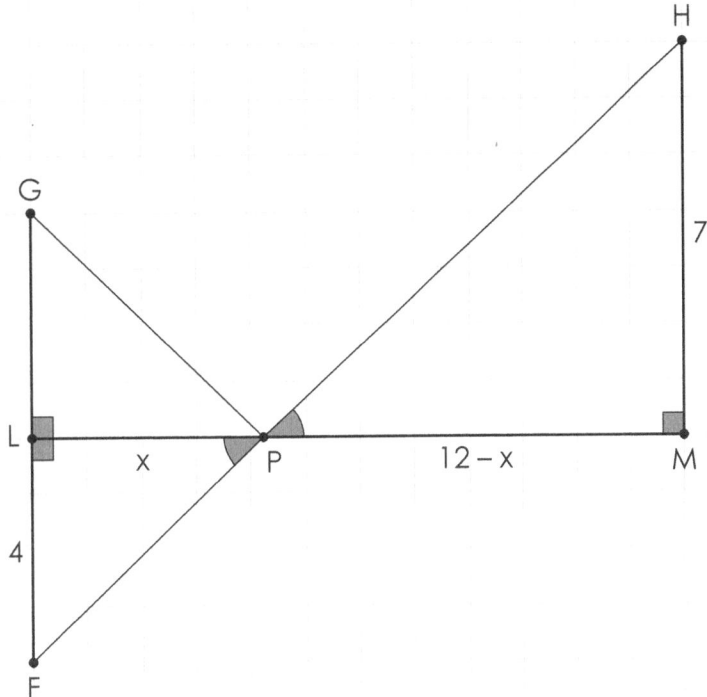

How far is $P$ from $L$? Because $\triangle FLP \sim \triangle HMP$ (why?), the corresponding side lengths have equal ratios:

$$\frac{x}{4} = \frac{12 - x}{7}$$

Students who know how to solve basic linear equations can use cross products and the distributive property to solve this. Others may guess and test.

$$7x = 4(12 - x)$$

$$7x = 48 - 4x$$

$$11x = 48$$

$$x = \frac{48}{11} = 4\frac{4}{11} \text{ or } 4.\overline{36}$$

# ALGEBRA CONNECTIONS

Stage 1 is about simplifying radicals. In elementary algebra, students learn to do this by finding perfect square factors of the radicand and applying properties of radicals. For example:

$$\sqrt{45} = \sqrt{9 \cdot 5} = \sqrt{9} \cdot \sqrt{5} = 3\sqrt{5}$$

Notice how Problem #1 helps learners visualize this process before they have been taught the "steps." Students will eventually apply the procedure to variables as well. You could challenge them to draw pictures to support a process like this:

$$\sqrt{5x^2} = \sqrt{x^2 \cdot 5} = \sqrt{x^2} \cdot \sqrt{5} = x\sqrt{5} \text{ (when } x \text{ is not negative)}$$

The Diving Deeper question in Problem #3 challenges students to create a formula for the location of $P$ that gives the shortest path. If $a$ and $b$ are the lengths shown, and $c$ is the length of $\overline{LM}$, the picture and the process might look something like this:

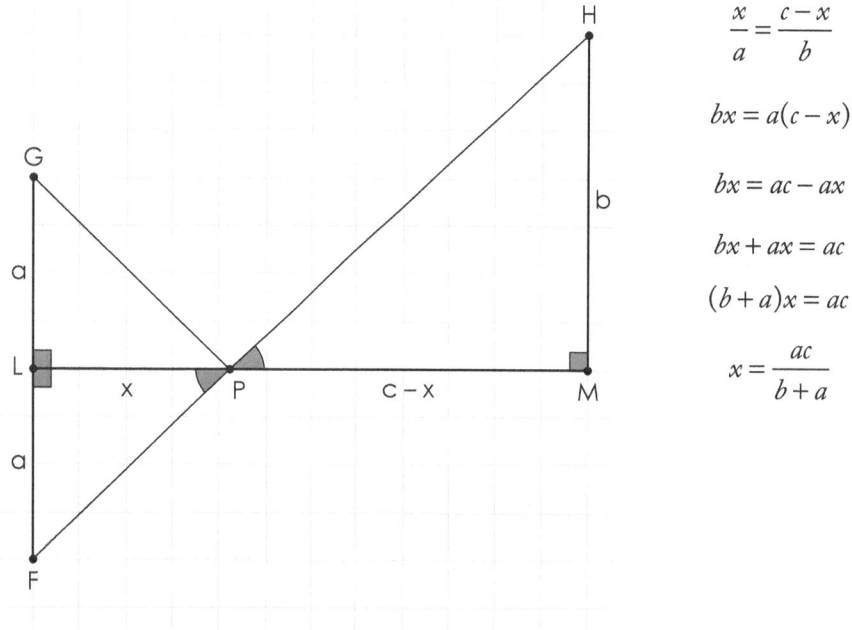

$$\frac{x}{a} = \frac{c-x}{b}$$

$$bx = a(c-x)$$

$$bx = ac - ax$$

$$bx + ax = ac$$

$$(b+a)x = ac$$

$$x = \frac{ac}{b+a}$$

Students may check that this works by substituting the values from the original problem. Notice how the algebraic process mirrors the calculations in Problem #3!

*I wonder* what the formula would be if you measured from $P$ to $M$ instead of $P$ to $L$?

# Exploration 10

## Twist and Shrink

Every polygon has a special point called its *centroid*. In this exploration, students will investigate surprising and beautiful connections between centroids and similar figures. One of my geometry classes accidentally discovered these connections a few years ago as part of another project. I decided to adapt the ideas as a task for my prealgebra students, and they have enjoyed exploring it while furthering their knowledge of similarity.

This is an open-ended project, and your students will need time, guidance, and a chance to collaborate to make the discoveries. Even then, they may not notice everything shown in the solutions. This is fine! Emphasize the concepts of your choice.

To get started, show your students how to find the *centroid* of a triangle. A *median* is a line segment joining the midpoint of one side of a triangle to the opposite vertex. In $\triangle ABC$, all three medians intersect at a single point. This is the *centroid*.

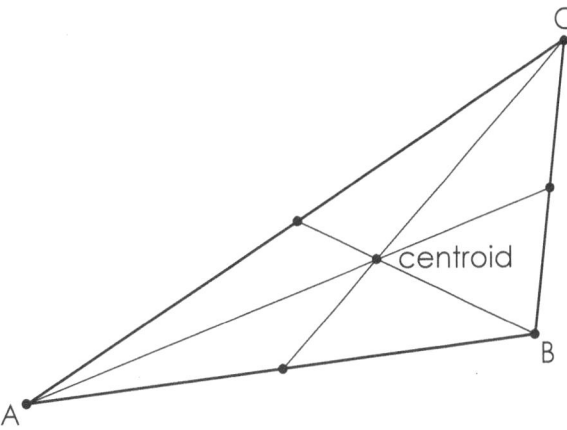

In high school geometry, students will learn that the medians of a triangle always intersect at one point. They will also discover that triangles have other types of centers such as *incenters*, *circumcenters*, and *orthocenters*. Some of your students may like to research and explore these.

Your students may be interested in knowing that the centroid of a triangle is also its *center of mass*. If you hang or balance a (uniformly weighted) triangle from this point, it will lie flat. Encourage them to experiment with this!

DOI: 10.4324/9781003232797-13

# STAGE 1

In Stage 1, students begin using centroids to create similar quadrilaterals (although they do not know this yet!). The process of finding the centroids can get a bit messy. I encourage kids to draw the medians very lightly. After they find a centroid, they can erase the medians so that the picture looks clean when they go on to the next centroid. I usually ask them to leave one set of medians in their drawing as a way of showing their thinking process.

The Solutions to this activity use "prime" notation to show corresponding points. For example, the point corresponding to $A$ is written "$A'$," which is read "$A$ prime." This naming system makes it possible to use the same letter for two different points so that it is easy to tell which points correspond.

## What You Will Need

» Graph paper.

## What Students Should Know

» Find the centroid of a triangle. (See the introduction to this exploration.)
» Plot points in a coordinate grid.
» Understand side and angle relationships in similar figures.

## What Students Will Learn

» Solidify and extend understanding of similarity, scale factors, and area factors.
» Begin to explore centroids of triangles and quadrilaterals.
» Make, test, and justify observations about complex shapes and processes.

# Problem #1

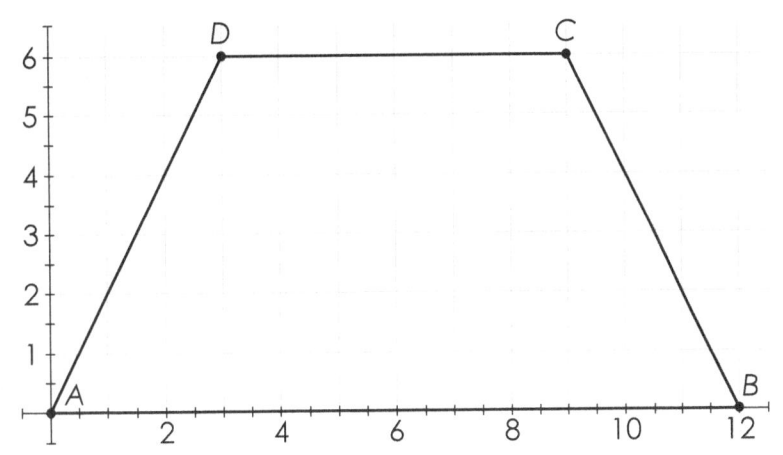

## Directions

- Plot the centroids of $\triangle BCD$, $\triangle ACD$, $\triangle ABD$, and $\triangle ABC$.
- Connect the centroids to make a new quadrilateral.
- Make and justify as many observations as you can.

## Testing the Waters

Solve the problem using this square.

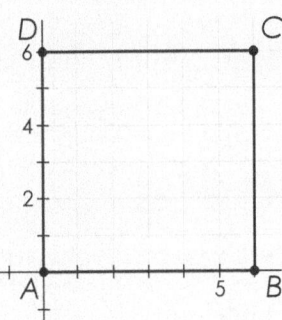

# CONVERSATION STARTERS FOR #1

*What do you notice? What do you wonder?*

*I notice* that I only have to draw two medians to locate a centroid.

*I wonder* what to focus my observations on?
There are many possibilities:
» congruence or similarity;
» measurements (angles, side lengths, areas);
» parallel or perpendicular segments;
» coordinates (the numbers in an order pair that tell you where a point is located);
» transformations (translations, rotations, or reflections); or
» connections between any of the above.

*I notice* that I could continue to make smaller and smaller trapezoids!

*I wonder* if I could continue the pattern outward to make larger trapezoids?

*I wonder* if I can predict the coordinates of a centroid without drawing medians?
Look carefully at the coordinates of its triangle!

*I wonder* how I would find the centroid of a quadrilateral?
One possibility: Use midpoints of opposite sides. Can you think of other ways?

## If Students Explore Centroids Of Quadrilaterals

*I notice* that both trapezoids have the same centroid.

*I wonder* if all of the trapezoids will have the same centroid if you continue the process?

# SOLUTIONS FOR #1

*Drawing (with some length and area measurements):*

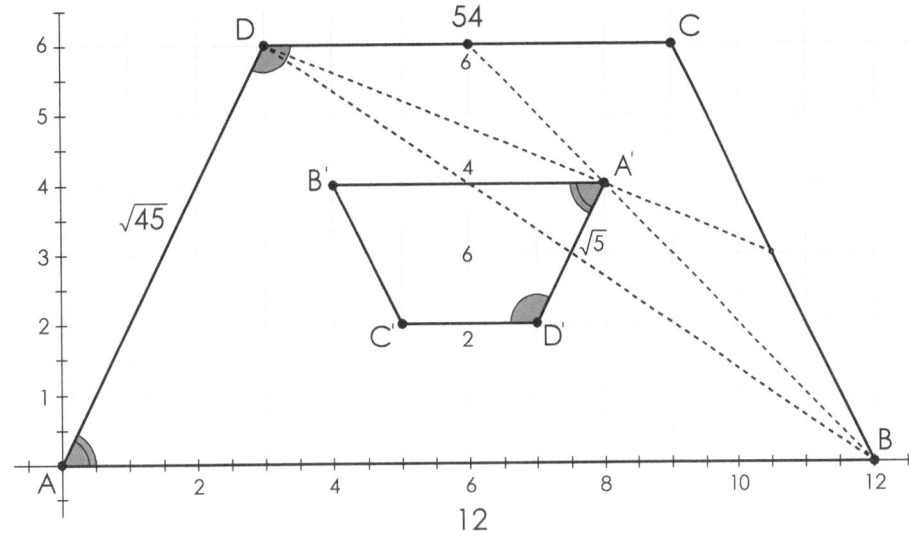

The light dotted lines show two medians that may be used to find $A'$ ( $A$ prime), which is the centroid of $\triangle BCD$.

*Some observations:*

» The new quadrilateral is similar to the original! *Justification:* Corresponding angles are congruent and corresponding lengths have the same ratio. $m\angle A = m\angle A' \approx 63°$ and $m\angle D = m\angle D' \approx 117°$. The lengths of corresponding sides have a ratio of 3. $12 \div 4 = 3$, $6 \div 2 = 3$, and $\sqrt{45} \div \sqrt{5} = 3$.

» The scale factor from the small trapezoid to the large one is 3. *Justification:* The scale factor is the side length ratio shown above. You could also see it using the grid to notice that 3 copies of each side of the small trapezoid fit into the corresponding side of the large trapezoid.

» The area factor from the small trapezoid to the large one is 9. *Justification:* The area of the large trapezoid (54 units²) is 9 times the area of the small one (6 units²).

» The area factor is the square of the scale factor ( $3^2 = 9$ ).

» Corresponding sides of the trapezoids are parallel. *Justification:* Segments that have the same steepness or slope are parallel. Count the horizontal and vertical distance from one endpoint to another. For example, you can get from $D'$ to $A'$ by going "right 1 and up 2." You can get from $A$ to $D$ by doing the same thing three times! (See the ramps in Stage 1 of Exploration 2 for a more detailed discussion.)

» The new trapezoid is inverted (rotated 180°).

» The $x$-coordinates of the centroids are the averages of the $x$-coordinates of their triangles' vertices. The same is true for the $y$-coordinates. For example, the coordinates of $A'$ $(8, 4)$ are averages of the coordinates of $B\,(12, 0)$, $C$ $(9, 6)$, and $D\,(3, 6)$:

$$\text{x-coordinate of } A' : \frac{12+9+3}{3} = 8$$

$$\text{y-coordinate of } A' : \frac{0+6+6}{3} = 4$$

# STAGE 2

In Stage 2, students experiment with a variety of quadrilaterals to see if their observations from Problem #1 still apply. They also begin to look more closely at centroids of quadrilaterals.

I usually start by asking students to guess how they would find the centroid of a quadrilateral (if they do not know yet). They may suggest two methods:

1. Join the midpoints of opposite sides. The intersection of the two segments is the centroid.
2. Find the centroid of a triangle and join it to the vertex that is not in the triangle. Repeat for another triangle. Find the intersection of the two segments. (This basically amounts to connecting $A$ to $A'$, $B$ to $B'$, etc.)

Both of these methods work! You may want to mention that, unlike triangles, the centroid of a quadrilateral is not necessarily its center of mass.

## What You Will Need

» Graph paper.

## What Students Should Know

» Understand Stage 1 of this exploration.

## What Students Will Learn

» Solidify and extend understanding of similarity, scale factors, and area factors.
» Find the centroid of a quadrilateral.
» Make, test, and justify conjectures about complex shapes and processes.

# Problem #2

Vertices of three quadrilaterals:

1.  A (6, 0)    B (12, 12)   C (6, 18)   D (0, 12)
2.  A (0, 0)    B (24, 0)    C (30, 6)   D (18, 12)
3.  A (12, 0)   B (24, 18)   C (12, 6)   D (0, 18)

## Directions

- Make one or more *conjectures* about connecting tri-angles' centroids in quadrilaterals.
- Test your conjectures on the quadrilaterals above (or others of your choice).
- Explore centroids of the quadrilaterals.

A *conjecture* is an educated guess or prediction.

*Advanced Common Core Math Explorations: Ratios, Proportions, & Similarity* © Taylor & Francis.

# CONVERSATION STARTERS FOR #2

*What do you notice? What do you wonder?*

Depending on how much students discovered in Problem #1, you may still use some of its Conversation Starters here. Some additional ideas are shown below.

*I wonder* if I can predict which sides will be parallel just by looking at coordinates?

*I notice* that the second quadrilateral is the only one that does not have any symmetry.

*I notice* that the third quadrilateral is concave.

*I notice* that the smaller concave quadrilateral is not completely inside the original one.

*I notice* that every large quadrilateral has the same centroid as the small one.

*I notice* that the segments I draw to find the centroid of the large quadrilateral overlap the segments I draw to find the centroid of the small one.

*I notice* connections between parallel sides and corresponding angles.

*I wonder* if I can still use averages to find the centroids of the quadrilaterals?

# SOLUTIONS FOR #2

*Conjectures*: Students may conjecture that their observations from Problem #1 will (or will not) continue to hold for all quadrilaterals. They should use drawings to provide evidence.

*Drawings*: The light dotted segments show one way to find the centroids of the quadrilaterals.

1.

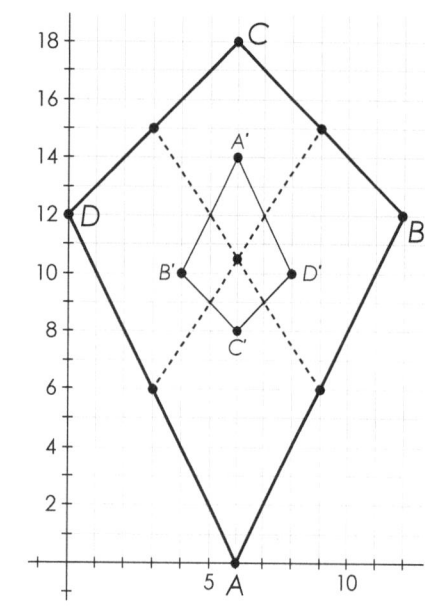

2.

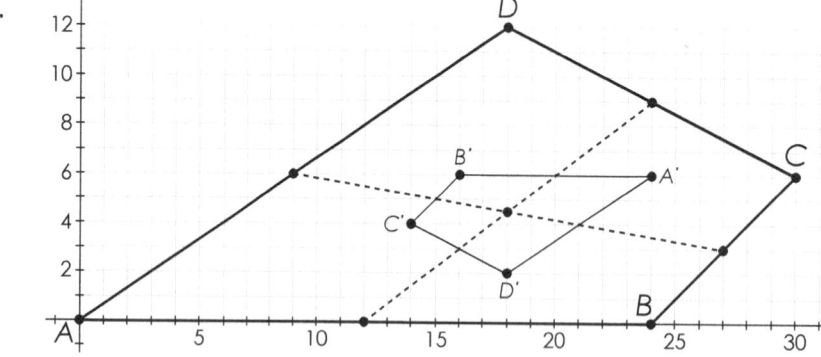

3.

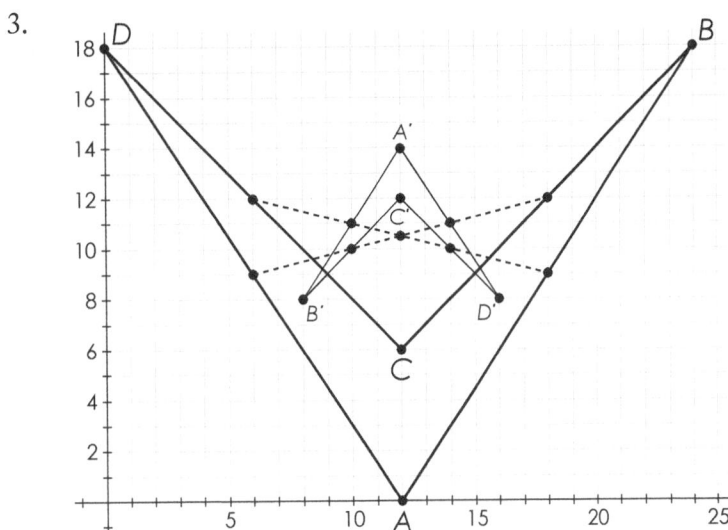

*More observations, conjectures, and connections:*

   » The conjectures from Problem #1 are still supported by these drawings.

   » Most of the small quadrilateral for the concave polygon lies outside the original figure. (So does the centroid!)

   » The quadrilateral's centroid may be found by averaging coordinates.

   » The similar quadrilaterals have the same centroid.

   » There are connections between parallel sides and congruent angles. If you extend some parallel sides to create a parallelogram, corresponding angles of the quadrilaterals are opposite angles in the parallelogram! For example:

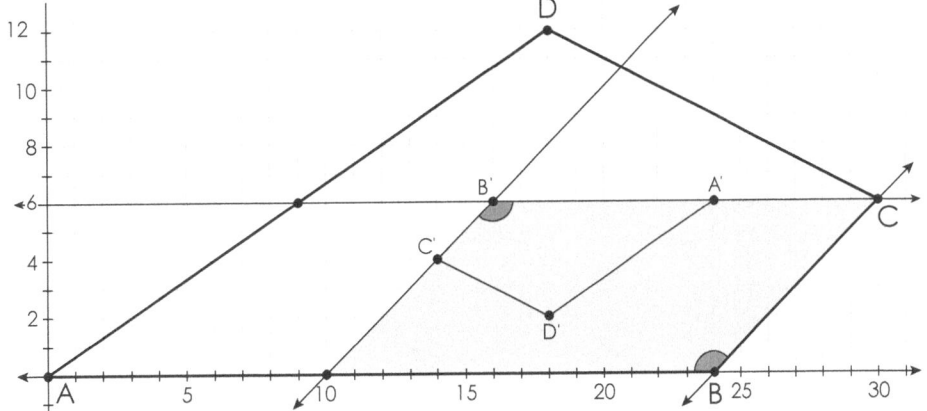

# STAGE 3

In Stage 3, students extend their knowledge from Stages 1 and 2 to pentagons. They will discover that most of their conjectures from quadrilaterals still apply. However, there are a couple of spots where they may be surprised or get stuck.

1.  They may assume that they should still use centroids of triangles to create the similar shapes. In fact, they need to use centroids of quadrilaterals.
2.  The scale factor and area factor change when you increase the number of sides!

I always let students struggle with these things for a while before giving too many hints. You never know—they may not need them!

## What You Will Need

»   Graph paper.

## What Students Should Know

»   Understand Stages 1 and 2 of this exploration.

## What Students Will Learn

»   Solidify and extend understanding of similarity, scale factors, and area factors.
»   Find the centroid of a pentagon (and possibly any polygon).
»   Generalize results about centroids and similar shapes to polygons with more sides.
»   Make, test, and justify conjectures about complex shapes and processes.

## Problem #3

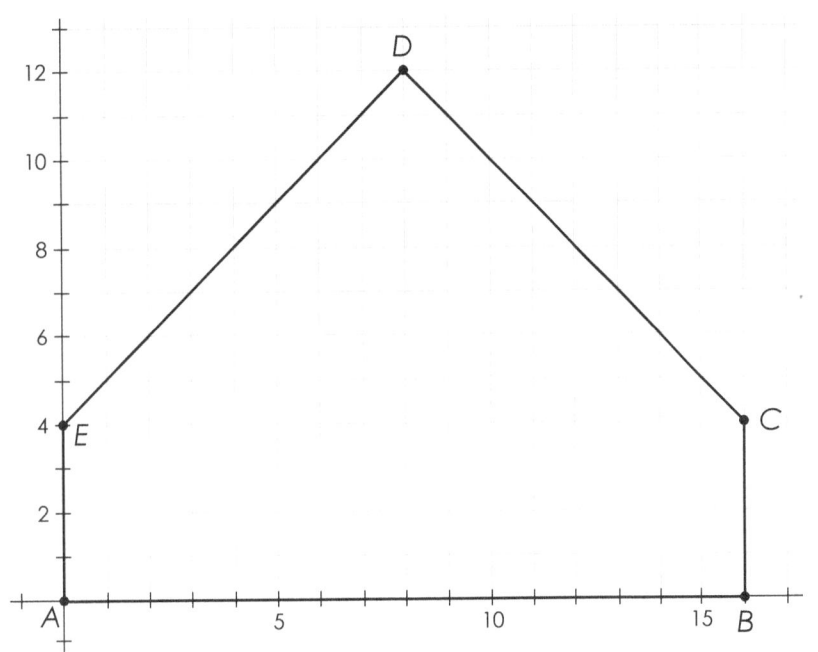

### Directions

- Make some conjectures about using centroids to create similar figures in pentagons.
- Test your conjectures on the pentagon above.

# CONVERSATION STARTERS FOR #3

*What do you notice? What do you wonder?*

Depending on what students discovered in Problems #1 and #2, you may still use some of their Conversation Starters here. Some additional ideas are included below.

*I wonder* if I can still use centroids to create similar shapes for pentagons?

*I notice* that using centroids of *triangles* does not work for the pentagon.

*I wonder* if the scale factor will still be 3 for pentagons?

*I wonder* how to find the centroid of a pentagon?

*I wonder* how to find centroids of polygons with more sides?

*I wonder* if there is a way to predict the scale factor from the number of sides?

*I wonder* what causes these patterns?

*I wonder* what happens if I apply this process to a triangle?

# SOLUTIONS FOR #3

Many students will probably try to create the inner pentagon using centroids of five triangles. They will discover that this does not work. What does work is to connect the centroids of the five *quadrilaterals BCDE, ACDE, ABDE, ABCE,* and *ABCD*!

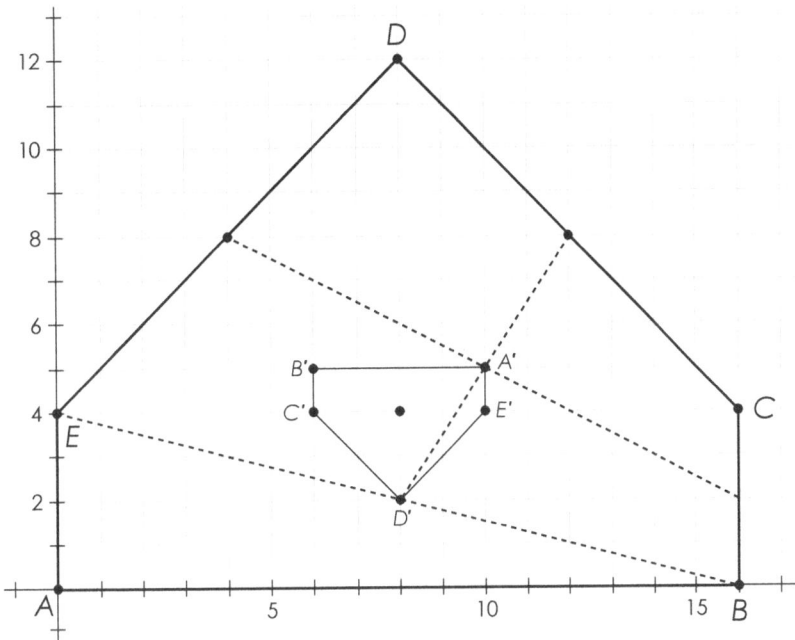

The light dotted lines show one way to find $A'$, the centroid of quadrilateral *BCDE*.

The scale factor is now 4 (or $\frac{1}{4}$), and the area factor is 16 (or $\frac{1}{16}$). Some students may conjecture (correctly!) that the whole number scale factor is always one less than the number of sides. Observations and conjectures that remain true:

- » The inner pentagon is similar to the original.
- » The inner pentagon has been rotated 180°.
- » Corresponding sides are parallel.
- » Each centroid's coordinates are the averages of the corresponding coordinates of its quadrilateral.
- » $\overline{AA'}$, $\overline{BB'}$, $\overline{CC'}$, $\overline{DD'}$, and $\overline{EE'}$ all intersect at the centroid of the pentagons.

*I wonder* how far the centroid is from the endpoints of these segments?

# ALGEBRA CONNECTIONS

Students who know about angle relationships involving parallel lines (transversals, alternate interior angles, etc.) may use them to explain why corresponding angles in the similar polygons are congruent. Some students may know these relationships informally from middle school work. High school geometry students may be able to prove the results formally.

Algebra students may extend the exploration by using general coordinates to prove their conjectures. By labeling the vertices of a quadrilateral as $A(x_1, y_1)$, $B(x_2, y_2)$, $C(x_3, y_3)$, and $D(x_4, y_4)$, they can use averages to find coordinates of the similar shapes. For example, the coordinates of $A'$ are:

$$\left( \frac{x_2 + x_3 + x_4}{3}, \frac{y_2 + y_3 + y_4}{3} \right)$$

With these expressions, they may prove that corresponding sides are parallel by showing that they have the same slope. For example, the slope of $\overline{A'B'}$ is:

$$\frac{\frac{y_1+y_3+y_4}{3} - \frac{y_2+y_3+y_4}{3}}{\frac{x_1+x_3+x_4}{3} - \frac{x_2+x_3+x_4}{3}} = \frac{\frac{y_1-y_2}{3}}{\frac{x_1-x_2}{3}} = \frac{y_1 - y_2}{x_1 - x_2},$$

which is also the slope of $\overline{AB}$! The expressions may seem a little frightening at first, but if you look at them piece by piece, they are not nearly as bad as they appear! Using the Pythagorean theorem (or the distance formula, which is the same thing) and properties of radicals, students may prove the scale factor relationship in a similar way.

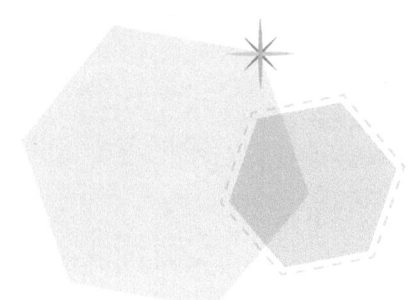

# References

Chapin, S., O'Connor, C., & Anderson, N. (2013). *Classroom discussions in math: A teacher's guide for using talk moves to support the common core and more* (3rd ed.). Sausalito, CA: Math Solutions.

Dweck, C. (2007). *Mindset: The new psychology of success.* New York, NY: Ballantine.

Johnsen, S., Ryser, G., & Assouline, S. (2014). *A teacher's guide to using the Common Core State Standards with mathematically gifted and advanced learners.* Waco, TX: Prufrock Press.

Johnsen, S., & Sheffield, L. (2013). *Using the Common Core State Standards for Mathematics with gifted and advanced learners.* Waco, TX: Prufrock Press.

Kilpatrick, J. (2001). *Adding it up: Helping children learn mathematics.* Washington, DC: National Academy Press.

National Governors Association Center for Best Practices, & Council of Chief State School Officers. (2010). *Common Core State Standards for Mathematics.* Washington, DC: Authors.

National Council of Teachers of Mathematics. (2000). *Principle and standards for school mathematics.* Reston, VA: Author.

Sheffield, L. J. (2003). *Extending the challenge in mathematics: Developing mathematical promise in K–8 students.* Thousand Oaks, CA: Corwin Press.

Smith, M., & Stein, M. (2011). *Five practices for orchestrating productive mathematics discussions.* Reston, VA: National Council of Teachers of Mathematics.

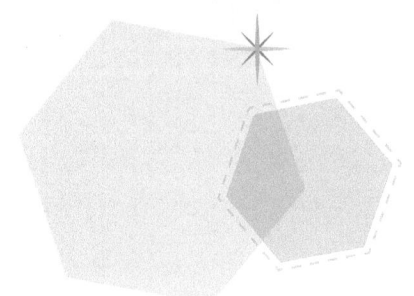

# About the Author

**Jerry Burkhart** has been teaching and learning math with gifted students in Minnesota for more than 20 years. He has degrees in physics, mathematics, and math education from University of Colorado, Boulder and Minnesota State University, Mankato. Jerry provides professional development for teachers and is a regular presenter at conferences on the topic of meeting gifted students' needs in mathematics.

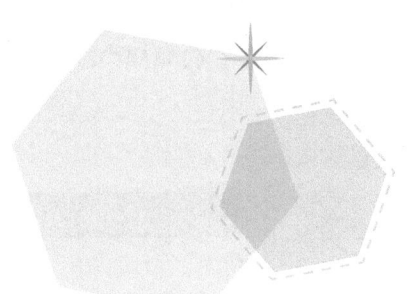

# Common Core State Standards Alignment

| Exploration | Common Core State Standards in Mathematics |
|---|---|
| **Exploration 1:** The Incredible Shrinking Universe | 5.NBT.A Understand the place value system. <br> 5.MD.A Convert like measurements within a given measurement system. <br> 5.MD.B Represent and interpret data. <br> 6.RP.A Understand ratio concepts and use ratio reasoning to solve problems. <br> 7.G.A Draw construct, and describe geometrical figures and describe the relationships between them. |
| **Exploration 2:** Ramps, Paints, and Hot Air Balloons | 5.OA.B Analyze patterns and relationships. <br> 5.G.A Graph points on the coordinate plane to solve real-world and mathematical problems. <br> 6.RP.A Understand ratio concepts and use ratio reasoning to solve problems. <br> 6.EE.C Represent and analyze quantitative relationships between dependent and independent variables. <br> 7.RP.A Analyze proportional relationships and use them to solve real-world and mathematical problems. <br> 8.EE.B Understand the connections between proportional relationships, lines, and linear equations. |
| **Exploration 3:** Gear Up! | 5.OA.B Analyze patterns and relationships. <br> 5.G.A Graph points on the coordinate plane to solve real-world and mathematical problems. <br> 6.RP.A Understand ratio concepts and use ratio reasoning to solve problems. <br> 6.EE.A Apply and extend previous understandings of arithmetic to algebraic expressions. <br> 7.RP.A Analyze proportional relationships and use them to solve real-world and mathematical problems. <br> 7.G.B Solve real-life and mathematical problems involving angle measure, area, surface area, and volume. |

| Exploration | Common Core State Standards in Mathematics |
|---|---|
| **Exploration 4:** Perplexing Percentages | 6.RP.A Understand ratio concepts and use ratio reasoning to solve problems.<br>6.EE.A Apply and extend previous understandings of arithmetic to algebraic expressions.<br>6.EE.C Represent and analyze quantitative relationships between dependent and independent variables.<br>7.RP.A Analyze proportional relationships and use them to solve real-world and mathematical problems.<br>7.EE.A Use properties of operations to generate equivalent expressions. |
| **Exploration 5:** Scaling a Tower | 6.EE.A Apply and extend previous understandings of arithmetic to algebraic expressions.<br>7.G.A Draw construct, and describe geometrical figures and describe the relationships between them.<br>8.EE.C Analyze and solve linear equations and pairs of simultaneous linear equations. |
| **Exploration 6:** Keep It in Proportion | 5.NF.B Apply and extend previous understandings of multiplication and division.<br>6.EE.A Apply and extend previous understandings of arithmetic to algebraic expressions.<br>7.RP.A Analyze proportional relationships and use them to solve real-world and mathematical problems. |
| **Exploration 7:** Grab Bag | 6.RP.A Understand ratio concepts and use ratio reasoning to solve problems.<br>6.EE.B Reason about and solve one-variable equations and inequalities.<br>6.EE.C Represent and analyze quantitative relationships between dependent and independent variables.<br>7.RP.A Analyze proportional relationships and use them to solve real-world and mathematical problems.<br>7.EE.B Solve real-life and mathematical problems using numerical and algebraic expressions and equations. |
| **Exploration 8:** Expanding and Contracting | 6.G.A Solve real-world and mathematical problems involving area, surface area, and volume.<br>7.G.A Draw construct, and describe geometrical figures and describe the relationships between them.<br>8.NS.A Know that there are numbers that are not rational, and approximate them by rational numbers. |

| Exploration | Common Core State Standards in Mathematics |
| --- | --- |
| **Exploration 9:** <br> Pythagorean Connections | 7.G.A Draw construct, and describe geometrical figures and describe the relationships between them. <br> 8.G.B Understand and apply the Pythagorean Theorem. <br> 8.EE.A Work with radicals and integer exponents. <br> 8.EE.C Analyze and solve linear equations and pairs of simultaneous linear equations. <br> 8.NS.A Know that there are numbers that are not rational, and approximate them by rational numbers. |
| **Exploration 10:** <br> Twist and Shrink | 5.G.A Graph points on the coordinate plane to solve real-world and mathematical problems. <br> 6.G.A Solve real-world and mathematical problems involving area, surface area, and volume. <br> 7.G.A Draw construct, and describe geometrical figures and describe the relationships between them. <br> 8.G.A Understand congruence and similarity using physical models, transparencies, or geometry software. |